G.I. DAYS

G.I. DAYS

AN ANTHOLOGY OF MILITARY LIFE

EDITED BY

MARY SENTER

MILLTOWN PRESS
Everett, WA

Printed in the United States of America
10 9 8 7 6 5 4 3 2 1

Cover Image: "A 19-Year-Old Marine in Vietnam,"
Danella Sydow, Acrylic on Canvas, 18 x 24 inches, ©2021.

Published by Milltown Press
Designed by Mary Senter

Library of Congress Control Number: 2023942070

ISBN: 979-8-9870418-1-9

This project is supported, in part, by a grant from the
Washington State Arts Commission.

WASHINGTON STATE
ARTS COMMISSION

Milltown Press
Everett, WA
www.milltownpress.com

Contents

Introduction

It hadn't occurred to me to curate an anthology of military writing until I was applying for a grant for military-connected artists. One of the questions in the application was "Assuming you were funded by this grant, how would you bring the benefits forward into your community?" As it was the first grant to which I'd applied, I didn't realize there was a "give-back" component to these things. I couldn't imagine how receiving a grant to focus on my writing would benefit the community, and then I realized I was ideally situated to give back to the military writing community. I had founded Milltown Press in 2021 with the intention of publishing one carefully curated title per year and I had yet to plan a project for 2023.

I proposed that I would publish an anthology of military writing, providing military-connected writers the opportunity to be published and share their stories. As soon as I submitted the grant application, I began to fear actually receiving the grant and having to make good on my commitment to publish an anthology with its enormous expenditure of time and personal resources. Oof! I knew I could do it, but it was daunting, nevertheless.

I received notification that I had been awarded the grant the day before the Association of Writers and Writing Programs (AWP) Conference was set to begin in Seattle. I'd been on the fence about attending AWP, but with the grant money, there was no reason not to. I began my 4-week grant period the next day by attending the conference with the intention of attending all of the panels on publishing, anthologies, and military writing I could find, in addition to panels that would benefit my personal writing projects.

The second panel that I attended on the first day—when I was still getting my bearings—was called "Disrupting the Veteran Writer Stereotype: More Than Males, More Than War." As I sat in the packed room and looked around at those in attendance and listened to the stories of the panel members and attendees, I knew these were the people whose stories I wanted to hear.

These were women who served on the front lines in combat only to find later that their service was swept under the rug, women who served during a time when wives wore dresses to serve their husbands dinner, LGBTQ folks who struggled to find their place in a military that didn't want or accept them, and others who just didn't fit the mold.

Military writing as a genre, has traditionally been dominated by stories of war written by white males. Not that these stories aren't important and compelling, but I wanted to hear the military stories of women, people of color, spouses and children, and people from marginalized groups, that are not necessarily about battle, but are about the everyday struggles that are unique to military life.

With new enthusiasm, I put out the call, not knowing if anyone would submit to an anthology by a tiny, unknown press, or if the call would reach its intended audience.

Fortunately, by the end of the submission period, I had received a surprisingly large body of wonderful work and I was saddened to have to turn away some great pieces that didn't quite fit the theme of the project.

My goal was to publish 30 writers, with a fairly even mix of poetry and prose, and that is what this volume contains.

I am thrilled to be able to present such a diversity of contributors and content in this collection. From historical pieces to humor, this anthology delivers something for everyone.

It is truly my honor to be able to put this lovely collection of work into the world. This project that was meant to "give back" to the writing community, actually gave back to me. In exchange for the labor I invested into this book, I have received an abundance of joy.

I'd like to give thanks to the Washington State Arts Commission for funding through the Wellness, Arts & the Military Self-directed Arts Practice Grant, without which this project never would have been conceived. I give a world of gratitude to Kevin Arnold, Bruce Sydow, and Danella Sydow for their incredibly generous donations to the project and to all of the contributors and others who provided support and encouragement along the way.

I'm grateful to everyone who entrusted me with their work. Thank you for sharing your truths (and fictions).

Mary Senter
September 3, 2023
Everett, WA

Refugee Heart

Brendan Praniewicz

I know that tear gas smells like apples,
and apples, to Kurds, symbolize love.
Tears streamed down my student's face
because her mother and sister, poisoned
by nefarious gases of Saddam Hussein.
I know that my heart stops beating
when a Chaldean boy shouts in class,
"I watched ISIS behead my father,"
or when a Somali girl buries her head
because terrorists kidnapped her brother,
or when a young refugee boy from Iraq
sprints home in the middle of my class,
and he weeps into his mother's chest,
"Momma, I'll never learn this language."
"My son, one day, you'll master English."
Today, he's a better writer than I am.

I'm no refugee, but roads run through me.
My veins, tangled in various time zones.
My father served a lifetime in the Army.
A product of countries, borders, zip codes,
I've run my finger against the Berlin Wall.
I know freedom flows in westbound lanes.
Texas sunsets burn in lavender and pink.
I know a couple of words in broken Greek.
I can recognize an Appalachian accent,
and ocean waves, like people, vary daily,
and you can breathe clouds in Himalayas,
and if you stand in Saint Peter's Square
and shake sunflower seeds in your palm,
a pigeon will perch on stretched fingers.

I'm no refugee, but pain runs through me.
My father has been present my whole life,
and never had a present moment with me,
and addiction isn't always sour and dark,
but sweet and tart like tangy orange sherbet.
I know that my father has bipolar disorder,
and that anger always terrifies my soul.
And when my rage begins to boil over,
I will reach a place where I will laugh.
And I know, once I'm closer to laughing,
I'm closer to letting all of my anger go.
My father taught me—laughter empowers.

I know that my heart stops beating,
when I venture home to Pittsburgh
and witness wrinkles on my mother
and my father shuffles a little slower.
I know that I'm going to die one day
—everyone is going to die one day.
I know that poverty sounds like rain
tapping on a trailer's broken windows,
and how to warm hands on an oven,
when it's the only heat that you have.
I know that my heart stops beating,
when driving home on a cold night,
I see an old woman on dirty blankets
as she falls asleep on a soiled sidewalk.
When I see suffering, I see my mother.

I know that all of people that you love
will hurt you more than anyone else,
and all of those people you despise,
sometimes, they pleasantly surprise.

One day, a hundred years from now,
our lovely actors, writers, and singers,
no one will remember any of them.
One day, we'll forget Shakespeare too.
All the borders, we bleed and die for,
will all sink into dirt and nothingness.
I know that only one thing is certain
and it's that this life has no certainty.
I know I must embrace others tighter,
and that I need to love so much harder.

Everywhere and Nowhere

Patty Somlo

I'VE NEVER BEEN SURE WHICH ONE to use. So, I alternate. Asked by a stranger where I'm from, I sometimes answer, *nowhere*. Other times, in response to the same question, I say the opposite. *Everywhere* becomes my response.

I always add the same phrase, though, to my perplexing answer. "I'm a military brat," I explain.

Everywhere has made me a citizen of the world. As a young girl, I danced the hula in Honolulu's Kapiolani Park, as part of the King Kamehameha Day celebrations. Years later, while attending high school in Frankfurt, Germany, I enjoyed long, thin bockwurst sausages dipped in spicy mustard, bought from small sidewalk stands.

Nowhere means that for decades none of the rundown apartments I rented or the cities where they were found felt like home. That word made it possible to say goodbye to friends and jobs, colleges and homes, rarely missing anything or anyone once I had gone.

Many years into adulthood, I met a woman named Lisa, who'd had an upbringing like mine, and she reminded me of much I had forgotten. Lisa recalled having to pause whatever she was doing, walking outside or waiting to pay for a purchase in the PX (post exchange), because the American flag was being lowered at that moment, and everything on the base needed to stop. Like me, she'd grown accustomed to picking up and leaving one home and moving across country or the world to a different one she'd never seen before. As I'd done, she learned how to make friends quickly and easily let them go.

Here and there, I met others who'd lived this same disjointed life. I started having contests with newfound acquaintances who were veterans of military service childhoods, to see which of us had attended the most schools. I often won.

Everywhere and *nowhere,* I gradually came to learn, prepared me for life and didn't. *Everywhere* made it possible to walk into a room full of strangers, and in moments start a conversation, and maybe a relationship. *Nowhere* left me incapable of forming deep intimate ties or settling down, buying furniture and maybe even a set of matching plates. *Everywhere* and *nowhere* ran into each other like bumper cars, making me argue with myself about where I wished to be and what I wanted.

Some of the clearest childhood images that have stayed with me are of my father coming home from, or leaving on, a trip. In the returning home memory, he is lifting presents he's brought me back from Japan out of his olive-green flight bag. We are sitting in our narrow duplex living room, across the street from the officers' swimming pool at Hickam Air Base, Hawaii, the most idyllic location since I loved to swim. Though decades have passed, and the presents are gone, I can picture them—zories with fat black velvet straps above woven bamboo bottoms; a geisha doll in a red brocade kimono, enclosed in a glass and bamboo case; and pink plastic chopsticks to wear in my hair.

I can also picture my father in his poufy olive-green jacket out on the flight line, where he's driven the car. After he steps out with his matching bag, my mother slides behind the wheel and drives my sister and me home.

In a moving introduction to the book *Military Brats:*

Legacies of Childhood Inside the Fortress, by Mary Edwards Wertsch, the late author Pat Conroy addresses the suffering endured by military children. He points out that while our military parents are usually honored for their service to the country, we offspring never are. Conroy suggests a parade should be held in which military brats march past their service moms and dads sitting in the stands. As their children pass, the parents should jump to their feet and salute them.

For the longest time, I considered my nomadic childhood as a lucky start to life. What other kid got to spend three years in Hawaii riding waves on Oahu's Windward Coast, walking barefoot on the sand, and learning to dance the hula and play the ukulele? Who else got to live in Germany, learning to ski down Austrian slopes, riding horses in Berchtesgaden, or visiting Versailles in Paris and the Acropolis in Athens, Greece?

I loved that life, or so I thought. After high school, I entered the School of International Service at American University, hoping to one day join the foreign service as a diplomat and work at posts around the world. But the coursework was too rigorous for me, who instead of spending time preparing for a career desperately wanted to be having fun. Several years and different academic majors later, I earned a bachelor's degree in art education. That didn't stop me, though.

It didn't stop me from dropping everything at a moment's notice and moving on, mimicking the life I had as a child. I quit jobs, often before securing another one. I changed homes, relocating across town or taking up residence in a different city, state or country. In my relationships, the same pattern could be found. Friends, lovers, and acquaintances slipped through my life, one minute here and the next minute gone.

I held on to the fierce belief that I could create the life I yearned for simply by moving. Hadn't I done that when I was young? In each place, I made friends, found my way to and from school, and learned to love something about the strange town or Air Force base where we'd come. I had a successful track record, didn't I?

One weekend, though, that hard-earned faith drained out. The latest in a string of brief romantic relationships fizzled. For the first time, something became clear. No matter how many times I met Mr. Right, he would, before long, turn into Mr. Wrong. Every job would one day become unbearable. I would never make a home, regardless of where I landed or how many apartments or cities I tried.

A friend suggested short-term therapy. I nodded, and later that morning picked up the phone.

Once a week, I sat in a straight-backed wooden chair across from my therapist, Barbara, and went on a journey to a place I'd never been. That place, my past, was the terrain I learned long ago as a military brat to forget, not grieving the friends I lost, the beaches I loved, or the home I once adored. I grieved the life that for so many years I'd celebrated, realizing the price I paid and what I gave up, for always being ready to move on.

Bit by bit, I began to picture a different life, one that included marriage, long-lasting love, a home with comfortable furniture, and friends I wouldn't toss aside. I met a man named Richard, and we married, moved from San Francisco to Portland, Oregon, where we bought a Victorian house. For several years, we worked with a contractor to restore the worn-down Queen Anne Victorian cottage, transforming it into a beautiful home. On weekends, Richard and I scoured antique malls to furnish the place, mimicking bed and breakfasts we loved.

With a wonderful husband by my side, a landscape photographer who lived to travel, *everywhere* had new meaning in my life. We visited nearly all the Hawaiian islands, snorkeling behind neon-colored fish above pink and green coral in Maui, marveling at rainbows in the midst of showers over Kauai's Waimea Canyon, and awestruck by the big-wave surfers on Oahu's North Shore. In Glacier National Park, we clapped loudly to warn bears on trails to lakes and waterfalls and later watched a moose dunk her head in a lake, as she ate calmly, at the end of a hike in Canada's Waterton Lakes National Park.

While we loved Oregon's snow-covered peaks, wild and scenic rivers, sparsely visited trails and endless lakes, after twelve years in Portland we'd grown tired of the nine-plus months of annual rain. We bought a mid-century modern ranch house in Northern California wine country and started to renovate.

A year into our life in a small city known for good weather and even better wine, I found something I'd quietly sought for decades. Yes, I had carried the word *nowhere* on my shoulders nearly all my life, but *nowhere* was crowded out by the place that now felt like home. *Nowhere* had faded, just as the shade of dark blue vanishes from a pair of denim jeans washed way too much.

That same sense of home comforted my beloved husband in his final days, when he could no longer hike up steep mountain trails to photograph. Instead, he stepped slowly out to the back yard, where he snapped close-up shots of the beautiful old roses we'd inherited from the previous owner.

Before his death from cancer, he sat with me every afternoon, in a black wicker chair underneath the fig tree, and we reminisced about our life. In place of planning our next travel adventure, Richard and I relived favorite times in places we'd already enjoyed. We also pondered what we couldn't know —how I would go on without him and what lay in store for him after he left this life.

Five years to the day from when Richard was first diagnosed with stage four cancer, I arrived on the Island of Kauai in the midst of a blustery downpour. Driving the compact rental car to the one-bedroom condo I'd reserved, I could barely see the truck ahead of me through the rain and water being flung against the windshield.

Though I was staying on a part of the island Richard and I had visited many times, we hadn't been back for over a decade. In addition, whenever we travelled, Richard always drove.

I surprised myself, though. After I crossed the bridge over the Wailua River, I realized that I knew where I was. A moment later, I took a right turn off the highway. When I then pulled into a driveway, a short distance down the road, I spotted the sign for the condo complex, letting me know I had arrived.

The following evening, I walked the short distance to the Coconut Marketplace, just past the hotel where Richard and I had stayed on our first visit to Kauai. A small crowd was already gathered on the grass-covered hill overlooking a small stage. I walked over to an empty spot on a concrete bench in front, and sat down.

After the band played a familiar Hawaiian song, two young girls walked to the stage and started dancing. They both held *ipus*, brown calabash gourds, the same ones I played when dancing the hula on the neighboring island of Oahu. The lovely girls were about the same age I was then. As I watched them dance, I let the tears fall, feeling the circle of life making its inevitable journey once again.

Veteran's Day, 2020

Kait Walser

Thought of you this morning,
my oracle. Shimmering: a flame
and my sparkling slip-ons. A call.
A text. Tulip and pomegranate.

Reach for the rosy vortex. A nebula
for ideas, ready to be born of that
mindset: suspended between daydream
and boots on the ground.

My father didn't speak to me
for a whole stretch of holidays.
And yet, Wednesday, he called:
Happy Veteran's Day.

It was his way of recognizing those five years
together: when I negotiated rides, hiked
up my too-big BDUs, tucked bleached
strands back in a bun and saluted the brass.

Got praised at truck stops as if I were a real
sailor, not a Sea Cadet. Just a kid in polyester
parades. Sipping post-grinder Gatorade.
Hot chocolate after the shooting range.

When the line went quiet, I'd think back to him: half
a world away, lining up at his base's phone booth just
to say *goodnight* right before bedtime. Nowadays, we're all
scuttlebutt and chit chat. Laughter and *love you* every day.

Sometimes, we don't have the words.
Sometimes, we have everything else.
The right ASVAB score. Two languages
or more. Company Honor Cadet. A heart

willing to plunge from the top of the high dive.
Just to try. Sometimes, we can see ourselves
in the sparkle of a loved one's eyes. We might
recognize when their path just isn't quite right.

Sometimes, I still wonder: what if? I wonder
if Jupiter's red eye still blazes through the soft
clouds of my aura. My father's pride: a chevron
on either shoulder, stretching out to fresh wings.

Before the Blood Drive, April 2020

Kait Walser

My love stirs lentils, beans, fresh cilantro. We sit.
My brother calls. Another rough day and now gunshots

on the quiet Pennsylvania hill where we grew up. He wants one *to defend
the home*—the house he wants to leave, didn't mean to come back to—

and anyway, he has another place lined up. He says I know nothing
about guns, but my ego begs to remind him of my trainings as a teen

on military bases. My marksmanship ribbon and the shooting ranges.
It's no use with his brimstone tone. *You've never had a gun aimed at you,*

fired at you, so you wouldn't know. And besides, he says, back when he sold,
he had two. *Just ask Dad—I was way more hotheaded way back then.*

Him, our archetypical Aries. How my Libra rising cast his fists in spackle shrines
while repairing the drywall. I picture the quaint scene of Maine our dad painted

for me, where he lived *before you two were on the scene.* EMT school and lobsters
in concrete pools. When he showed up eager to learn how to save people,

how to heal. Phlebotomy and bartending. Finally, years after our dad's ink sleeves
were at capacity, after the needles stopped sending his vagus nerve in a tizzy,

he gave blood again. We have the same type, which doesn't surprise me.
My brother calls to apologize. And he means it, I think. He thinks so at least.

I tell him *I have so much faith in you and your future* and I mean it, at least I mean for
him to hear *your future* and believe in it.

Communications and Misunderstandings

Kathleen Isaac-Luke

IN 1966, FRESH OUT OF NURSING school, I joined the US Army Nurse Corps as a first lieutenant. After completing officer training at Fort Sam Houston, I was stationed at the 5th General Hospital in Bad Cannstatt near Stuttgart, Germany. The base housed a complex of clinics and a hospital of nearly 400 beds. When you walked out of the main gate, you were in the historic town of Bad Cannstatt with its half-timbered houses and mineral baths. Within easy walking distance were local shops and restaurants or Gasthäuser.

After I became acclimated to my assignment on the pediatric unit, I was determined to learn the language of the country which would be my home for 2 years. Through the hospital pharmacist, who had arrived earlier than I did, I was introduced to *Frau* Schneider, a tutor who had earlier translated for the US Consulate in Frankfurt.

Twice a week I took the streetcar to her flat in Stuttgart for my lessons in beginning German language. She was very strict about my pronunciation and vocabulary and drilled me on both while Debi, her terrier, nipped at my ankles. One of my most vivid memories about my lessons were her dog's frequent disruptions and my tutor's admonitions of *"Debi, Lass das!"* That phrase was imprinted in my memory, though fortunately I never had occasion to use it.

After several months of study, I felt ready to venture off the base armed with my new repertoire of German phrases. I was in need of a haircut and one of the German secretaries on my unit recommended a salon that was a short walk away. I called and made an appointment for a haircut and shampoo with a stylist named Charlie.

Feeling very confident about my German vocabulary, on the day of my appointment, I marched into the salon and announced *"Ich habe eine verabredung mit Charlie."* The entire staff of the salon burst into laughter and called out for Charlie, who was the only male stylist in the salon. He appeared from a back room and joined the rest of the staff in laughter.

I could feel my face turning crimson and wondered what had provoked such amusement. Was it my accent? In confusion I turned to the receptionist who seemed approachable. *"Was ist so komisch?"* I asked hesitantly. She replied in perfect English, *"Eine verabredung* is a date— between a man and a woman." I didn't share that *Frau* Schneider had taught me otherwise.

Charlie directed me to his station and began to trim my hair. Periodically, everyone in the shop, including the clients, broke into laughter. Soon, I joined in the levity, although I did not think it nearly as funny as they did.

At my next session with *Frau* Schneider, I shared with her what had happened when I tried to practice my German skills. She shook her head sternly and opined, "They were so foolish!" Eventually, through her tutelage and the kindness of many German acquaintances, I learned to navigate the language. And, although Charlie gave me a very good haircut, I never visited that salon again.

Haole (We're Moving)

Noelle Dennard

When I was ten years old, other girls scolded me
For how I said Ms. Furukawa's name,
So I tried to speak like they did,

And they only laughed at me.
So I just stopped saying certain things.
I couldn't even say *haole* right.

I know we're not the victims.
I see how the military eats up all the housing prices,
Chasing people out to Las Vegas.

I hear them complain about living in a "shoebox,"
And how their free house
Is a dump compared to what they had in Texas.

I see the military pollute the water supply,
Pollute the blue oceans, fail to *mālama* the ʻaina,
But even so, I was just a kid

Being told I was a colonizer and a foreigner,
Not sure how to belong
In the space between "here" and "home."

I don't think I have a better home.
So when the coffee shop lady explains what taro is,
And the farmers market lady explains what a mango is,

I feel stamped with the label of *haole*—
Foreigner
One who doesn't belong here.

But if not here, where do I belong?
Do fifteen years of life mean nothing?
I never felt tan enough,

And I spoke with the accent of my parents, not my peers,
And I ate haole food, because my ancestry is far from here…
But this is still my home. I promise it is. I plead that it is.

I'm only caught between two somethings—
Colonizer and immigrant? Resident and tourist?
Military and local? I don't know

But I fear I was never meant to be here
Despite the kindness and the smiles
So *aloha ʻoe*, I'm going home to somewhere…

No Longer with Us

Benjamin B. White

Coast Guard Group Moriches
Was where the local Reserve Unit—
         Comprised of many
         New York City
         First Responders—
Would hold their drills,

And in October 2001,
On the first duty weekend
After 9-11,
There were many
         Coast Guard Reservists
Who didn't answer
Roll Call.

Halfway Round the World

Irene T. Winslow

Halfway round the world
I discovered
The love I lost
The day I left
The security of my home

Halfway round the world
I discovered
The love
I left
Halfway round the world

Halfway round the world
I discovered
That love
Left
Halfway round the world
Is with me
Still
Halfway round the world

Chain of Command

Nicole Powers

WE, AS MILITARY BRATS, PROUDLY SERVE our country, patriotic, unrecognized and silent, at attention as the dependents of the American military. We grow up in a cocoon, protected, insulated and regimented. Our lives are bound by the military dynamic of order and obedience. The only difference to set one brat apart from another is our father's rank.

While growing up, military brats deal with many things that require strict discipline and adherence to schedule and protocol. In some people's minds, it might seem as if there would be no time for a brat to take a deep breath and relax. It's not that a military brat has no time to be care-free; it's just a little more challenging to be a kid in that environment. I should know; for many years this was the confines of my life.

Once meals, school, and chores were completed, my greatest escape was to the forest's comforting embrace. It was while walking as the wind blows that I happened upon some abandoned barracks, cut down the middle lengthwise and lined up in rows and columns at parade rest, just like the soldiers they once housed.

I blew in and out of those half rooms, rustling the shaky pencil-thin walls, and disturbing any critters that might have thought of nesting in whatever corner they found comforting. My thoughts had found heaven and the perfect home, so I took possession and made it mine. I don't know if providence had chosen this discovery for me, but I was claiming it. It belonged to me.

A half barrack room became the great wall of

Hannibal (no reference to any historical event or location; I just liked the sound, Hannibal's Wall, when I named it) where I could defend against and conquer invading forces; another served as counsel chambers for secret negotiations; and one roomette was the grandstand from where I gave all my awe-inspiring speeches.

Each half room had a different purpose in the plays of my imagination. I spent hours and then days, furnishing and arranging my haven with the rocks, sticks and leaves I carried in from the forest. I had no need to plunder. All that I needed was bountiful. I even made accommodations for any guests that might wander by and guests did come, unwelcome as they were.

The first M.P. (Military Police) startled me when he banged on the door with his baton.

"Hey you, little girl, what do you think you're doing?" he yelled at me.

Without answering I jumped to attention, dropped my head, eyes front, and tried not to eyeball him. M.P.s hate it when you eyeball them. That is why I jumped out of my skin when the second one grabbed my shoulder from behind. Now I had a dilemma. Should I stand and defend hearth and home, fighting to the death, or surrender to the greater force?

All choices of what I should do were answered for me when the senior M.P. demanded, "Name and rank, soldier."

No way could I answer that question without condemning myself and my dad to the gates of military purgatory. Every brat has already painfully learn—if the kid gets into trouble, dad will pay the piper and you know what runs downhill when you step in it.

I took the only option I had; I cut and ran.

Delimiter

Ricardo Ruiz

02:31
The infinite blink of the clock
holds me in perpetual motion:
transfixed by the hypnotic rhythm:
I stare into each second: needing
to escape the neat tidy beat for
the ambiguity of a min : utes length:

02:32
I am walking through a memory:
:Boots on ground, you must walk the land
to own it: I was told: I still march:

02:32
Each footstep different and
each footstep the same with
no ownership
of now insight:

02:33
This memory owns me: I live in one
set of steps until the delimiters
light bursts and dust engulfs my body:

This memory owns me: I live in one
set of steps until the explosives
light bursts and dust engulfs my body:

Time and time again: I will it to
stop: release me from its spell as the
colon appears and then leaves a trail I : follow:

02:35
I work to reprogram the clock
I put tape over the colon
I:ve worked with a therapist
and taken the meds: Yet
I:ve not found a way to turn off the delimiter.
I want to close the separation
of what is and what was:

02:38

For the Good Doctor, When the Loss Was His Own

Michelle DeRose

He recognized every strep-carrying bump
filmed milky white at the back of throats
and withstood 365 days of triage,
deciding whose dangling limbs came first
in a war no one thanked him for sewing up.

He once scratched the throat of his toddler,
strung heels up, to free the airway
of meat chewed gray.
                              He looked
inside colons to find cancer and through
blood and skin for fractures.
                              He removed
warts, irrigated ears, drew blood,
and counted white cells.
                              He prescribed
four hundred drugs for two hundred diseases,
and told countless adult children
that their fathers died peacefully
or not.

                              But he needed a newspaper
to recall the date on the day
the middle name of his dead son
was more than he could remember.

Snapshot

Aliza Dube

YOU ARE AT ASHLEY'S SECOND CHILD'S first birthday party. It is a half-hearted affair, in the lull of the pandemic, during those lazy days of June when everyone still had hope that maybe this would be it, maybe life would be normal again now. The men are still overseas and will be for another four months or so. Ashley is still not taking it well. Your baby is sitting in your lap, in a Buzz Lightyear tank top intended for a child double his age. You kiss him on the crown of his head, because there are still moments, at random times, when you are over-taken by the wonder of him.

"I'm just so over that at this point," Ashley huffs, looking at you. Her child, the birthday boy, has been stumbling about in the same urine-soaked diaper since you arrived. "How long until eight again?"

Eight o'clock. The children's bed time. The appointed time when you can drag the pong table out of hiding. The hour of popped bottles, shot glasses. You forget some-times, how shockingly young Ashley and the other moms are until moments like this. Ashley became alcohol legal while she was pregnant with this last baby. She is only twenty-two. What had you been doing at that age? Who had you been? You are caught up in brief recollections of a ghost of a girl, sipping Jager in the passenger seats of some fling's car. These memories are something Ashley will never have. This is a privilege, you think, that you had time to know what it is like to be free.

The party is Pinterest perfect. There are golden bubble balloons taped to the kitchen wall that read "Happy

Birthday." The cake is special ordered, made by a caterer, the white icing as flawless as new fallen snow with a Dr. Seuss character lodged in the top. In pictures, this will look ideal. In the photo albums, this will be a beautiful memory. The child won't remember what it was really like. He'll never know that his mother spent his first birthday counting down the minutes until he was unconscious. He'll look at these pictures some day and say, *god, somebody really did love me.*

"Are you sure it looks OK?" Ashley is asking Mollie from the kitchen. She's fishing and you know it. She just wants someone to compliment her decor. But you can't really look down on that. Isn't that what you were all doing with your Facebook posts? You're all just looking for someone to assure you that yes, you are a good mom. Yes, you are doing everything right. No, don't worry, you are not fucking them up too badly, because sometimes, you're not really sure.

It is a little over a third of the way through your husband's deployment. You have all but given up on grocery shopping. You've had rice and chickpeas for dinner consistently for the past five nights without even batting an eyelash. You are trying to save money; you are trying to lose the baby weight. You imagine yourself squirreling away the stray dollars and change you are saving in a coffee can like the children from *Shameless*. You are saving. You are trying to claw your way back to Paris. You are kept awake at night by dreams of *arrondissements*, cobblestoned streets, pastel *macarons* melting between your teeth like spun sugar. You dream of smoking on fire escapes, the fairy lights of the Eiffel Tower glimmering in the distance. You have an irrevocable longing to walk the streets of a city where, if you listen closely, you can hear the echo of the Fitzgeralds laughing like children, before it all went wrong, as though it never would. You want to walk past cafes where Victor Hugo once spent his mornings scribbling philosophy and lacing his coffee with CBD oil. You have an aching to stand somewhere profound, to trace the footsteps of titans. You cannot do that here, not in Kansas. There is no poetry here.

But, in your heart, you know this really has nothing to do with France at all. It is about getting back to the self you were before all this. A person you haven't been for years, a girl who looked out at her future and saw no horizons, no boundaries. She is a girl who perhaps never existed at all.

You know, in your heart, you will never get there.

But you have lost the baby weight. You are back in your black crop top, with the neckline that plummets almost to your sternum. Your bleached hair cascades down to the small of your back. You can almost recognize yourself now, even if you are dressed for a night club and not a child's birthday party. For a blip of a moment, you can sit in your own skin without chafing. Maybe eight o'clock is all you are really waiting on too.

Elizabeth's husband is the only man at this party. He sits next to her on the couch but he is looking at you, in the way men have always looked at you, smiling and hungry like a wolf from a children's fairy tale.

Your wife is sitting right next to you, you want to tell him. *There she is, heavily dropping hints about how much she wants another kid, while you sit here undressing me with your eyes.* You know better than anyone, they will never have that second kid.

Later that night, Elizabeth's husband will bring

their three-year-old daughter home so Elizabeth can stay for the after party. Later that night, Elizabeth will be drunk out of her skull, her head resting in your lap, your back pressed against the chain link fence in Ashley's backyard. Elizabeth is mousy, with a low ponytail that reaches down past her butt. Ashley had grabbed the ponytail jokingly once, wrapped it around her fist, pulled Elizabeth's head back with it.

"This is why I'll never cut it," Elizabeth had said. "James loves to do that."

The other girls had laughed. But you still saw nothing but a leash.

You can see your baby through the sliding glass door of Ashley's house. He is propped in a car seat, watching cartoons with the five-year-olds as though he belongs with them. You never really expected him to sleep alone. After all, you never really could either.

"Sometimes I think I should just divorce him," Elizabeth admits, breezily. Her husband cheats on her. Big surprise. Everyone has the same story around here it seems but you. A coyote calls in the distance. There aren't any stars here, not the way there are back home. This fact always has a habit of making you feel more alone than you actually are. Your fingers ache for a cigarette, like the one Elizabeth has glowing between her finger tips, the cigarette that you know you can't have. That's the thing about bad habits, you suppose; you never really know how much a part of you they are until you have to quit cold turkey. It's something akin to attempting to remove your own liver with a butter knife.

"Do you think I should leave him?" she asks again. You barely know this woman, but you have spent a lifetime studying men. You know they don't change any more than you do.

"Yeah," you say, bumming a single drag from her cigarette. "I really think you should."

Bartending The V.F.W.

Thomas Lambert

Lawrence, Kansas (1995)

There's not much to it:
pour the drafts,
clear the ashtrays,
mix well-drinks.

The Vietnam vets
favor Jack-n-Coke.
The WWII crowd
Scotch-and-Soda.
Our only Korean War vet
takes his brandy neat.
We keep a bottle of St. George
behind the bar for him.

An occasional floor sweep
exceeds expectations.
Christmas décor is tacked up
year round.
Campaign insignias
and service ballcaps
line a smoke-filled perimeter
that obscures all
but the exit signs.

It's Memorial Day,
and the Colonel delivers
his annual speech
honoring the sacrifice
of our fallen brothers.

Manny offers a solemn toast,
"To the lost."

Jason buys another round
of shots.

To remember them,
they drink.

They drink
and beat their chests
in solidarity.

They drink
with the awareness
that all
are at the mercy
of the universe.

They drink
to animate themselves
and tell the stories
that otherwise
remain buried
in the footlocker of experiences
for which they do not
possess the tools
to relate
to one another,
let alone
civilians outside
eager to thank us
for our service.

Underway, off New Inlet

H.V. Rhodes

Fʀᴏᴍ ᴛʜᴇ ᴅᴇᴄᴋ ᴏꜰ ᴛʜᴇ USS *Allagah,* Seaman Richard O'Malley was the first to spot the strange ship as the tops of her masts rose above the horizon. He called to the officer of the deck, "Sir, ship in sight, due south and closing."

The sailors all stopped their work and stared as the stranger approached. The appearance of a large cruiser this close to shore was a novelty, for the biggest ships of the blockading squadron usually patrolled far out to sea.

Next to Richard stood Seaman Stephen Whelan. "Will you look at that?" Stephen said. "Maybe she's got our mail aboard. I'd sure like to hear from my Nora."

Four months ago, the *Allagah* had departed the Navy Yard at Brooklyn. Since then, Stephen hadn't received a single letter. In the same time, Richard had three letters from Helen, which was three more than he expected.

The tedium of blockade duty was broken only when the *Allagah* fired her four guns at shadows passing in the night. By day, the thin strip of green and brown that marked the coast never varied. While the days passed unremarkably, and weeks turned to months, the sailors of the blockading fleet maintained a stranglehold on the Confederacy.

As the cruiser closed the distance at high speed, the *Allagah's* officer of the deck used the fleet signal book to identify the stranger by her flag hoist.

The *Allagah's* petty officer of the watch cried, "Stand by to render honors."

"She's making right at us," Stephen said, "almost

like she means to ram us. That can only mean they have mail for us."

"Maybe." Rather than face his friend, Richard looked down at the deck. After all this time, if Stephen heard from Nora now, it would probably be bad news.

At the last moment, the large ship steered off and presented her broadside. Her three masts towered over the *Allagah's* stubby pair. She was nearly double the *Allagah's* length, had a crew five times that of the *Allagah's* and mounted ten times as many guns.

At the petty officer of the watch's commands, the *Allagah's* sailors lined up at the rail, snapped to attention and then saluted.

From the deck of the cruiser, an officer pointed a speaking horn at the *Allagah*. "In pursuit…Rebel pirate…any contact?"

Sometime in recent weeks, although none of them knew exactly when nor where, a Confederate commerce raider had slipped by and captured a dozen ships off New York harbor. Then, the mystery ship had disappeared.

"No contact," the *Allagah's* officer of the deck replied.

"We return…Hampton Roads."

"Have you…any mail…for *Allagah*?"

Each of the men on deck listened closely.

"No mail."

"Understand…safe voyage."

The large ship moved away at a good speed, trailing a long wake and a plume of black smoke.

"I'd like to know why our mail is delayed," Stephen said. "Hope the bag didn't get dropped over the side someplace. I hate to think of Nora's letters lost at the bottom of a harbor."

"That's enough, men," the officer of the deck called.

"Turn to ship's work."

Stephen shook his head. "Very good for them. Sure wish we was going north."

"Can't be helped," Richard said. "Who needs 'em? We're the ones that's winning the war down here."

Battling the Nazis

Mitzi Dorton

He fought his way through at seventeen
With few accolades or medals,
A bit of shrapnel remaining in his head,
Quiet, then blasts out of nowhere,
Endless moments of machine guns

The comrade he spoke to moments ago,
Mouth leaking blood,
Became a terrible telegram to a parent,
A family,
Somewhere in America.

Thank God he wasn't a wooden cross left behind,
That his family would never see.
Down in a frozen ground foxhole.
On the front lines, fodder for the war
With an excellent cause

He came home,
But not without World War II calling out in the middle
   of the night.
He survived to wash his new car in the driveway,
Gazing at the plane in the air,
With flashbacks of parachutes and supplies.
Above the rainbow spray of the water hose,
His child came into focus
With aluminum tumblers and neighbor friends
On the front steps,
Mouths red with Kool-aide mustaches.

A Job That Paid

Jennifer McKeen Rodrigues

young but not knowing it

     I longed for him
     while he played in the desert

waiting for a hurried love I had
uncertain ways of knowing about

the advertising job
that left me crying in the air

after work I yearned for

the cut of fabric
creating dresses

sold online
to the highest bidder

     for him to return home
     for our life to begin

I waited with the azaleas
  the magnolias
    the dogwoods

an everblooming romance
that arched over the continents   the oceans

made haste by wedding plans
the oak mansion built to witness our life vows

a humid expression of our devotion
that grew more and more by distance and silence

when he roamed the desert
gun slung on back   I waited

for weekly phone calls
   optimistic emails
      cut off by mortar shells

it was a job that paid

we were all working for the money

his more dangerous
more exciting by design

we spoke of a honeymoon
St. Barths St. Maarten

forgetting to do my work
while the office chatted   I waited

with the frozen turkey dinner
   family quarrels
      exodus from my job

a state of being in waiting
so young but not knowing it

A Family of Service

Mary Gutierrez

Technically, I'm not in the military nor am I retired from the military. I've never deployed. I've never held a weapon in my hands during war time or any time for that matter. I've never worn a military uniform. I respect and honor those who have. And although I've done none of these things, I've served, just not in the traditional capacity that people are used to. I've served as a member of a military family. I don't know if people recognize the sacrifices that families make for those who serve. I'm not complaining. I just think it's important that it is brought to light.

I come from a family of generational military service. My father served in the Air Force and my brother served in the Air Force, both retired. My nephew is currently active in the Air Force. There are so many other family members that have served—Uncle Louis and Uncle Joe in the Navy, Grandpa Gutierrez in the Army, Uncle Charles in the Army, and a cousin in the Marines. Here's why I believe that counts, being from a military family, and why, once in a while we should get some love.

I am old enough to remember when my father went to Vietnam. I was young enough, however, that I didn't understand everything that was going on. All I knew was that we had to move literally across the country from Florida to California and more importantly, my father was going to war. I knew war was bad. I saw things on TV, ugly things, things beyond my young mind's comprehension. I didn't understand why we were at war. I didn't understand why people were not supporting our troops. Not supporting my dad.

Back then we didn't have computers, internet, Face-Time. We had letters. Letters that got lost or took a long time to get to their intended reader. And as a child, waiting to get that letter was difficult, painful. Not getting a letter from my dad when I thought I should have was also scary.

Even as a kid I couldn't help but go to the worst-case scenario, which was based on a combination of what I saw on the TV and overheard the adults talking about. Although Mom always assured me that it would get there, I carried a sick feeling in my stomach until it did. I spent most of my time feeling scared for my dad, myself, and our family. Just trying to survive. What would happen if Dad didn't come back? Where would we live? What would happen to us? I kept these things to myself. I now realize that this was a lot for a young kid to carry.

I remember that he was able to come home for a visit once. He wasn't there for long, but I was just happy to see him. I remember crying when I saw him. That is when I learned that you could cry when you're happy just like you cry when you're sad. I was so grateful when he finally came home to stay. I knew, even then, that many did not. He retired from the U.S. Air Force on Jan. 31, 1978, after 24 years of service. During his service, he earned three Distinguished Flying Crosses and 21 Air Medals. We were together for the next 50+ years until he made his transition.

My brother, my only brother, followed in my father's footsteps. He started with ROTC in high school and then he excelled in the Air Force, graduating at the top of his class, reaching top positions. It made my father and family proud. I'm grateful my father got to see this. My mother got to see some of his accomplishments as well. I saw the worry on her face however when there was a chance he might be going somewhere she didn't want him to go.

Everything is OK until you hear of the U.S. entering into war. Anyone who isn't scared by this has never had a family member serve. After 9/11/01 it was inevitable that we would go to war. How could we not? There was so much fear and hatred at that time. He didn't go right away, and he had been sent to other places when there were skirmishes going on, but this, this was different. We all knew it was different, military and nonmilitary people alike. I remember things clearly about 9/11 and later when he got sent to Afghanistan. Whether a child or an adult, the realization that someone you love will be put in harm's way, no matter what the reason, is frightening. There are so many uncertainties. Where are they going? Can they tell you? How long will they be gone? Can they stay in touch? The family member's perspective is so different than those who serve, but I'm sure some of the questions might be the same.

Family members carry a different load. A load that is carried in the heart and stomach. Similar to when my dad was in Vietnam, when you don't hear anything from your loved one, you walk around with a knot that never goes away and the continuous feeling that you're going to be ill. It was like that the whole time he was in Afghanistan. You're completely distracted from everyday life until they come home alive and safe. And gratefully, that he did. He continues to serve; he now works for the Veteran's Administration.

I have to tell you about my mother and her brother, my Uncle Charles. His nickname was "Buddy" and my mother loved him. I never met him. He was in the Army and was killed in training maneuvers in Germany. We have a portrait of him that hung in every house we ever lived in. It still hangs in our family home, now my sister's home. There wasn't always a lot of discussion about Uncle Charles, but when there was that pain from his loss shrouded my mother's face. Her eyes would water, she would get quiet, and have a faraway look. She carried that heartbreak with her all her life, and I carried it as well because she was my mother. They are now buried in the same cemetery in Meridian, MS.

My nephew hasn't seen any war time and the protective aunt, the selfish aunt, hopes he never does. He's married now and expecting their first child. We know there is always that probability that he might get deployed if something were to happen, but I pray it never does. I'm not as resilient as I used to be, and my heart can't afford to be broken.

I don't expect anyone to ever thank me for my service. I'm good with that. I wouldn't want them to thank me. I do want people to remember that it takes a special kind of person to serve and a special kind of family to stay home and wait not knowing what the outcome might be. We all make sacrifices as part of a military family.

My Father Flew in Blimps

Louise Moises

In World War II, my father flew in blimps,
great grey whales of the air
floating on the hum of small engines.

His ship dipped above the house
where my mother lived,
before she was my mother
or he my father.

She ran out onto the porch wildly
waving, in her innocence, believed
he could hear her voice
throwing love into the sky,
where the bulbous Navy blimp
teetered above the house.

The First Seal

Bill Cushing

Navy Lieutenant William Barker Cushing, "Will" informally, became the first in his family to learn of his brother Lon's death at the Battle of Gettysburg.

He was at Naval Headquarters in Washington, D.C. when couriers delivered word of the battle. Will was a deck officer for the USS *Minnesota*, a blockade ship cruising the mouth of the Roanoke River in North Carolina, when a messenger interrupted the briefings to deliver the news. Two years earlier, in July of 1861, he affirmed his dedication to Lon by proclaiming, "If the Rebels should kill him, I should become a fiend."

In many ways, that is exactly what he did—just not the way he planned.

Upon hearing of the death, Will requested a transfer from the Navy to the Army, hoping for command of the 4th Artillery to honor the brother he loved so dearly. Fortunately for the Union, and in many ways history, the Army didn't cooperate since Will achieved unlikely feats of daring throughout his Naval career.

Two months after his brother's death, while standing mid-watch, he took the *Minnesota's* helm to chase down and collide with a Confederate blockade runner, but that only marked one incident among many.

In his youth, he was sturdy—almost burly, headstrong, and hasty, qualities he didn't shed when entering the Naval Academy in 1857 at 14, an even earlier age than his brother. Now only 19 years old, he was the youngest lieutenant in the Navy, which might not have seemed possible given how his military life had started. His family used political connections to get him into Annapolis hoping to instill personal discipline in the young man.

Despite the Cushing involvement in the political realm of the United States, Will had no interest in such things. In fact, he generally disdained politics, once writing, "I went up to the new house of representative and while there listened to enough nonsense to last me a year."

As a midshipman, however, his love for the Navy took a fierce hold over him. The youngest of four brothers, Will intended "to see every nook and corner of this little world that is to be seen." In his journals, he once proclaimed, "I had rather be an officer on board a man-of-war than the President of the United States."

Not being political in any real sense, he never looked at the big picture. As a student, Will was not very serious, especially if he saw no practical end to a course of study, and at the time, Annapolis focused primarily on a Classical education. He excelled in classes focusing on the tactical or military aspects of his chosen career such as gunnery, navigation, or seamanship classes. English, history, or physical science held no interest. He saw no purpose in them.

His behavior was equally comical and rebellious. Will was known for playing pranks on both staff and fellow classmates such as propping buckets of water over the tops of doors to douse unsuspecting entrants. At times, he fabricated a grappling hook and rope to snag jugs of tea steeping on mess hall window sills meant for faculty, later lowering the emptied container back to its ledge.

Then there was his Moses imitation.

"Platoon leader Cushing," barked his company commander, "what in the name of all that is holy were you thinking?"

"About what, sir?" Will replied, a quizzical look plastered on his face.

"You marched your unit right into College Creek today."

"I was acting guidon bearer for the company, sir; we were in forward march."

The upper-class officer's frustration boiled over.

"And you know damned good and well that the roadway bends to the east toward the river there!" he shouted, his face reddening and his nostrils flaring from anger.

"Of course, sir," Will said, deadpan. "But we never got the order to 'wheel right,' so I continued marching forward—as commanded."

"And you know that the officer commanding your unit stutters!" Slamming his fist against the desk, the ranting midshipman shook it in pain. Will's brigade commander, sitting to the side, worked to suppress a smile as the other continued his tirade. "He couldn't get the words out in time; it is your duty to accommodate such a situation when it arises."

"Sorry, sir," Will said, feigning ignorance. "I was simply following orders."

"Very well: now follow this. You just earned ten demerits."

"Thank you, sir," Will replied, standing, saluting in mocking precision, and leaving the room, again garnering points against him for actions born out of his attitude.

Will held the distinction of accumulating excessive demerits every year, changing his conduct only as the count began approaching a number that would lead to expulsion. The following year, he would go right back to his normal patterns of mischief.

While he might have gotten away with many of these acts, what didn't help was a tendency to question authority figures for whom he had no respect, including faculty.

He was nearly expelled from the academy for dousing a Spanish instructor with a bucket of cold water as the man was on his way out to try and woo some women in town. After Will drew an insulting sketch of the same instructor, in a class he wasn't doing very well in anyway, the academy's superintendent—a very conservative traditionalist preferring the previous practice of culling officers from the "right" families—saw an opportunity. He coerced Will into signing his formal resignation from Annapolis on March 23, 1861.

The Civil War began less than two weeks later, which gave Will a chance for redemption. He traveled to Washington, D.C. to plead the case for his reinstatement to a cousin who had a connection to Gideon Welles, the Secretary of the Navy.

Once he'd returned to the fold, Will defined his career with incontestable bravery that was almost manic. Once aboard his first ship, he wrote to a cousin: "Where there is danger in battle, there will I be, for I will gain a name in this war." The war offered numerous chances for him to do so.

Since the existing warships of the Navy were designed for oceangoing battles, they drew too much draft to enter most of the coastal rivers. The Union's strategy was to use its naval forces to blockade supplies and weapons coming to the Confederacy. Will found himself stationed at the mouth of the Cape Fear River in North Carolina since its twisting rivers and hidden inlets allowed Rebel blockade runners and foreign vessels relative success in running contraband through the region.

Finding action wherever and whenever he could,

Will was credited with numerous captures. In May 1861, he had a hand in taking five schooners including the first prize of the war, a tobacco runner. By the time he was 19, he was given command of the USS *Ellis*, a gunboat. In one two-week period, he boarded and seized two merchant vessels, one holding cargo worth $75,000 and the next with an inventory valued at twice that—an impressive bounty that allowed him a certain leeway although his defiant nature did push limits.

By its nature, blockade duty was one that required patience, never Will's strength or even a minor attribute. Waiting wasn't in his character, so he often took things into his own hands. He'd sneak off to counter the boredom of the squadron's assigned task, making daring and risky assaults on Hatteras region forts.

On numerous occasions, his superior officers wanted to "bring him before the mast" for punishment after each incident.

The problem was that every time he strayed, he managed to accomplish some task that no one else seemed capable of. By the time word of what he'd done moved up the chain of command, his superiors were stymied from action by a newspaper article publicly touting the young officer's heroics. In one instance, the *Ellis* ran aground trying to return and had to be destroyed. An official Naval board of inquiry convened to investigate the loss of its ship. During the hearings, they learned that he had destroyed ten Confederate vessels along with an important processing plant. Faced with that information, even the admiral he served not only conceded his worth as a leader but wrote a letter commending "Lt. Cushing's coolness, courage, and conduct."

Will became an icon for personal bravery that

bordered on bravado, and readers loved the daring young Naval officer depicted in the news. While he didn't follow orders very well, he embodied the later military idea of adapt, improvise, and overcome.

And he is unquestionably remarkable in the naval history of the 19th century. Thus, he's acquired many labels as the first: the first commando, the original special ops sailor, the first SEAL. It became obvious the young officer could not contend with the monotony of blockade duty, so the admiralty decided to let him loose, provided he inform them what he was planning. That concession paid off handsomely, as the young lieutenant produced two noteworthy successes within months of each other in 1864.

In February, Will proposed taking a contingent of men to kidnap Brigadier General Louis Hebert, the regional Confederate commander. Once he secured permission, he took to the task with gusto. He began the mission by approaching the town from a direction that would fool those on shore that his contingent was moving down river and most likely friendly. He got to the residence only to discover Hebert wasn't there.

Still, he couldn't resist tweaking the general's nose a bit.

"My Dear General," Will wrote on a note card that he left propped against a candlestick on the dining room table. "I deeply regret that you were not at home when I called. Very respectfully, W.B. Cushing."

Even the Confederates admired that kind of brass, and several of them made note of the Navy's renegade.

Admiral Samuel Lee, head of the squadron, decided to give Will free rein to form his own command and hunt blockade runners. In three months, Will got another chance to prove his cunning when the

CSS *Raleigh* emerged from the river and attacked the Union fleet. Will immediately seized on a new idea: rather than defeating the *Raleigh,* he would capture it.

On a moonlit evening, Will and 15 volunteers boarded a 30-foot cutter to find the ironclad warship. To get closer to the vessel, he took a veering and crooked path to deceive Rebel guards on shore that he was retreating. Then, some passing clouds blocked the moon. He took advantage of the natural cover for the more direct route upriver.

Upon arrival, the team discovered the *Raleigh* had run aground and split its keel. It was mired in the muck of the river bank near Wilmington, North Carolina, making its capture moot. The trip appeared to have been wasted time. However, Will took the opportunity to observe and map the city's defenses, turning what might have disheartened others into an intelligence coup.

In addition, one of his crew captured a Confederate rider carrying mail. Going through the bag's contents, they found letters containing information about the size of a garrison in a nearby fort, an inventory of its supplies, and the deployment of its guns.

Even this wasn't enough to placate Will.

Whether bravery, audacity, or a bit of both, Will donned the courier's coat and hat and rode the horse to a general store. Using Confederate money from the mailbag, he bought food and supplies. This raid proved its worth in terms of information and material; its boldness convinced the young officer's superiors that they had access to a rare talent, a man who combined intelligence with cunning. As a result of his earlier adventures, they tagged Will as the candidate to confront and hopefully neutralize another Confederate ironclad, the CSS *Albemarle.*

Known to Union sailors as "the Confederates' Roanoke River bully," the *Albemarle* had established itself as the unrivaled power on the Roanoke River. After attacking and inflicting serious damage to four Union gunboats during one battle with only slight damage to itself, one senior commander of Union naval forces in the region conceded that "the ram has possession of the river." Its armor was so effectively constructed, that cannon balls actually shattered into pieces, sometimes causing added damage to the crews firing them as shrapnel from the rounds ricocheted. During another confrontation that lasted three hours against six Union battleships, all were damaged while the *Albermarle* left both undamaged and without suffering a single casualty.

The Union Navy tried several raids to disable the craft, but direct attacks were made difficult by the Confederate fortifications dotting the shores leading to its dock. Attempts to use moored torpedoes like floating mines failed as well because its armor extended down its entire gunwale. In August of 1864, a lure was set out in hope of ambushing the ship, which resulted in the capture of ten of the men lying in wait

After brainstorming several ideas—none of them feasible—the Navy realized it needed someone able to conceive a decisive and destructive response. The admiralty agreed that Will was the perfect person for the task even though they considered the feat impossible.

"Impossibilities are for the timid," Will countered after hearing the idea. Besides, for him it was personal. One of his closest friends was killed during one of the attacks by the seemingly invincible monster. When offered the mission, he promised, "I shall never rest until I have avenged his death."

Once the Navy offered him the challenge, little else needed to be done but to wait for him to hatch his plan. This certainly was Will's territory, for he took his time despite his personal animus after losing both his brother to the war and a close friend to that particular ship.

Now sitting in an office of the Naval Shipyard in Brooklyn, New York, he contemplated the project before him. Assessing the enemy vessel, Will scribbled notes about the *Albermarle*:

200' L

50' W

—extra wt/armor affects speed & maneuverability.

Early eyewitnesses reported that its deck held two 6.4-inch double-banded rifles along with a moveable cannon that could be shifted to fire from any of several different places. Then there was *Albermarle's* apparent invulnerability. Sailors engaged in earlier battles with it recalled cannonballs bouncing off its thick iron hide, and, according to available intelligence, it was four inches thick in places and protected by nearly a foot and a half of wood backing.

Will hoped to board and "take her alive" but knew such a feat was unlikely given the number of guards he expected to be protecting such a valued weapon. Barring success with that idea, his next option was to approach the much larger vessel with what seemed a boat so small it would be deemed unworthy of much attention. From there, he and his men could explode the charge underneath the *Albermarle* and, if not sink it, at least render it useless.

He paused, clenching the pencil between closed teeth, and thought. It was the ship's combination of fire power with its protective measures that became Will's priorities.

Will soon deduced that its weakness was, quite literally, the beast's underbelly. He would attack it from below the surface rather than butt heads with it directly in open waters. Success rested on getting close enough to inflict the damage.

"May work," Will muttered to himself. "The trick is stealth."

Leaving his office, he went out to the yards and gathered some shipwrights, taking them to the wharf. He had already seen two boats he wanted to use. He chose them since they operated on low pressure steam engines, making them the quietest potential attack craft.

"That's the lead boat," he pointed to the larger of the pair. "I want it outfitted with a hinged stem that can be mounted with a moveable spar. Make sure there is a halyard to raise and lower the boom."

He then visited the gunnery sheds to acquire a torpedo, a fairly new weapon but one that a Confederate submarine had used successfully about eight months before; it seemed a good starting point. He approached the chief gunner and challenged him to design a method of mounting a torpedo at one end of the lever that could be lowered into the water ahead of the attack boat while being exploded from the other end.

"Can you do that?" he asked the chief petty officer.

"Count on it, sir," the old salt replied, honored to be part of whatever was underway.

Because of the nature of his plan, Will would need volunteers as daring as himself. After sending word, 275 volunteers appeared on the first day. Like his older brother Lonnie, Will commanded respect from most who served with him. Although he was known for his earlier pranks, when it came to military matters, Will was a strict disciplinarian, but sailors revered him. Some even offered to forfeit any bounty for the chance to take part in one of his famed exploits. This meant that finding the 26 sailors he figured he'd need turned out to prove the easiest part of the whole plan.

His only initial requirement was to eliminate any married men from the list. The danger inherent in this mission seemed to offer little else than certain death. He let it be known that those wishing to sign on that they wouldn't know what they'd do until they were on their way. His only promise to them was "death, glory or promotion."

He would take 14 with him in the lead picket boat, his choice because, even when fully loaded with men, equipment, ammunition, and weapons, they would only draw four feet of water. The other dozen men would follow in another to lend a hand if needed.

The next step circled back to Will's first thought: how to sneak up on the enemy as silently as possible. Taking blankets and makeshift insulation, he instructed the crews how to wrap the steam engines to muffle noise, having them practice so that they could get it done quickly once they were approaching the target. As the men drilled to perfect that task, the chief called him over to show this proposed design.

"Here it is, sir," he held an 18-foot iron pole in both hands, clearly proud of his design. Pointing to one end, he continued. "We'll mount an explosive canister at this end with a mechanical blasting cap and then here, at the other end, we'll mount a trigger mechanism for you to detonate the whole thing."

"Thanks, chief." Will clapped him on the shoulder. "With luck and your ingenuity, we'll take this monster down."

While the crews prepared the small ships, he took time to visit his mother. He assumed he was not going to be returning from this mission. Because his assignment was so important, he took her on a carriage ride one day to tell her, swearing her to secrecy.

"Mother," he said, "I have undertaken a great project, and no soul must know of it until it is accomplished, but I must tell you for I need your prayers."

Her response? "I know you will do your duty."

Will returned to the shipyard, and he and his crew departed New York as October neared its end. Will hoped to take advantage of the natural cover provided by a waning moon. The two vessels made their way to the Roanoke, entering the mouth of the river on the evening of the 27th. The night was rainy and cold but, a few hours after midnight, proved fortuitous as a translucent veil of rain combined with a foggy mist hugging the water's surface to impair the vision of any lookouts. The attack team approached its target at about three in the morning.

Standing on the foredeck of the lead craft, Will spotted the outline of the *Albermarle* at its moorings. He could smell the thick smell of burning tar from sentry bonfires and knew it wouldn't be long before the Confederates—.

"Who goes there?" challenged one of the sentries.

"We'll soon let you know!" Will shouted back. The other Union sailors joined in the fun, shouting derisive comments and jokes. The Confederates on watch hesitated a bit watching this small and apparently lightly-armed boat coming toward them.

He cast caution overboard and turned to his helmsman to shout, "Ahead fast!"

"Damn Yankees're out of their minds," a soldier on shore yelled to his comrades as he prepared his musket to shoot at the charging cutter.

Standing high on the prow of the lead boat, Will swung the spar-and-torpedo fixture into position. Bullets whistled past the sailors as a Rebel volley began. At the last moment, he noticed that a log boom circled his target—a floating defensive barrier that fires from shore had revealed.

"Double back," he ordered his helmsman, who turned the launch around. Will thought if they could build up enough steam, they should be able to vault the slippery obstacle. On the return approach, having gained momentum, Will's craft slid over the slime-covered barrier.

With bullets going past and around him from both the docks and the ship, Will, completely calm under fire, began working the improvised mechanism of his torpedo. He could hear the *Albermarle's* deck-mounted cannons being fixed and turned toward him and his crew, looking up for a second to stare directly down the barrel of one of them.

Fumbling with the torpedo's release line, Will battled darkness, wet, and enemy fire to focus on the task at hand. He ignored the clang of metal on metal as the Rebel crew slammed the cascable and sealed the ammunition in the loaded cannon. He heard the loud report of the fired artillery piece and could smell the sulphureous gunpowder. However, because the launch rode low in the water, the canister flew over the crew's heads, whistling as it went by. Almost simultaneously, Will pulled on the triggering line with his right hand, detonating the torpedo beneath the Albermarle with a muffled thud. The explosion tore a gaping hole in the ship's bottom and sent a tall column of water around its hull and into the air.

Will's plan seemed to have worked, but his own boat

was severely damaged as well. He yelled a final command.

"Every man save himself!" Will and the others dove into cold water while being pelted with grapeshot from rifles and muskets. Once in the river, Will swam until exhausted, crawling ashore. He could hear several Confederate search parties pass close by but was only discovered by a slave who'd been ordered to seek out any Union sailors he might happen upon. Naturally, the man was not as enthusiastic in completing his task. When Will saw the black man standing over him, he asked what had happened back at the berth of the *Albermarle.*

"She is dead gone sunk," the slave told him. "And they will hang you if they catch you."

Reinvigorated by the success of his mission, Will drove forward and back to his squadron. He spent the rest of the next day trying to get back to his squadron, hugging the shoreline and wading through water made prickly by the autumn cold whenever he wasn't hiding in the marshes.

When he was rescued by a pilot vessel sent out as a search party, he was worn out, haggard, and scratched by thorns, biers, and branches. As bad as he looked, there was no mistaking they'd found Lt. William Barker Cushing.

"Perhaps," he told the men who'd found him, "I was unobserved because of the mud that covered me and made me blend into the earth."

On October 30, 1864, he began writing up his official action report, which opened with this statement: "I have the honor to report that the rebel ironclad *Albermarle* is at the bottom of the Roanoke River."

His idea had worked. The torpedo punctured the *Albermarle,* sending the feared warship deep in the muddy bottom and leaving little chance of salvage.

In a letter to his mother, he proudly proclaimed how "that ex-midshipman, ex-master's mate, hair-brained scapegrace" had succeeded. Once shamed at the Naval Academy, he had become the national hero of the Navy.

President Abraham Lincoln called for a proclamation of thanks from Congress, even meeting him in person. Navy Secretary Gideon Welles observed that "young Cushing was the hero of the War." One American writer who understood sea captains dedicated to the completion of their mission, Herman Melville, noted the event by writing the poem "At the Cannon's Mouth."

After the war ended, Will continued moving up the ranks. In 1867, he was attached to the Pacific Squadron as executive officer of the USS *Lancaster*, the flagship of the Pacific Fleet. It was part of a flotilla making ports of call throughout Africa and Asia. While stationed in China, he proved his wild streak hadn't mellowed when he and a fellow crew member broke into the Forbidden City at a time when it was—quite literally—forbidden. Yet, even that breach of international protocol turned out well for him. When Imperial guards caught the two Americans, they were more concerned about being punished for dereliction of duty than they were of the foreign trespassers. The Chinese soldiers shooed them away.

Upon his return to the states in 1873, Will was transferred back to the East Coast for a patrol of Central American and the Caribbean under the auspices of the Monroe Doctrine. Nearly a decade after his best-known heroics of the Civil War, he was named commanding officer of the USS *Wyoming*, a sloop of war. While off the coast of Panama, Will received word that the Spanish had captured the USS *Virginius*, an American vessel in Cuban waters.

Although he had neither authorization nor orders, Will ordered his ship into the Caribbean to intercede. He understood that it was easier to seek forgiveness than permission.

At the time, the military governor of the island was General Juan Burriel, a man of such brutality he was nicknamed the Butcher. Living up to this nickname, Burriel began executing the crew of the *Virginius* as pirates.

When Will arrived and demanded an audience, Burriel refused, underestimating what he saw as a brash young American. Will had been promoted to full commander at 28—again, the youngest Naval officer ever given that rank, but he was not about to be brushed aside. His reply, devoid of any diplomacy, was a dispatch sent ashore to the general and governor. "If any more prisoners are executed," it read, "I shall open fire on the Governor's palace."

Reading the message, Burriel looked out in the harbor to see the *Wyoming* being brought broadside and setting battle stations, its crew fixing six heavy cannon barrels in his direction. Burriel understood the weight of the action and succumbed to a face-to-face. Once ashore, Burriel approached Will and held out his hand.

Will responded with a cold stare, his hands clasped behind his back, letting the general know there were to be no courtesies practiced here. Burriel was dealing with a lion, one whose pride was still keen and whose claws could still damage an opponent.

Facing Burriel down on that dock at Santiago de Cuba with such obvious abruptness and incivility spoke volumes. The military governor chafed at the rude response, but the obvious decisiveness of the young captain forced his hand. In an attempt to establish control over the situation, Burriel began lecturing him on "American interference in sovereign affairs" but was interrupted by Will, now gripping his sidearm while speaking.

"Are you going to shoot any more of the crew of the *Virginius*? Give me an answer; I will not be insulted again."

Burriel tried to buy time.

"*Perdoname*," Burriel told Will, softening his tone. "I truly wish I could accommodate your wishes. Unfortunately," he shrugged his shoulders, "any authorization would have to come from Havana."

"In that case, sir," Will replied, "I must request that all women and children be removed from the city, for these are not my wishes. They are my demands."

As he pivoted to return to the *Wyoming*, Will's message couldn't have been made any clearer. He would give the command to open fire on the city and was obviously just the sort of man to carry out any threat he made.

Burriel agreed at once to halt the executions. He realized he was faced with a warrior who wasn't about to grant any quarter.

That confrontation marked Will's final military victory of record. Unfortunately, he was a gladiator in need of enemies and combat, closer to his ancestors as a Celtic berserker than a naval officer, but failing health finally forced him ashore.

Will's daring recklessness put him on shaky ground throughout his life, but his impact on Naval thinking proved that effective was better than epic. Naval battle to that point had been grand and expansive; Will preferred working in a narrowly focused and personal manner. His influence on Naval warfare was so great that five naval vessels have been named in his honor, all of them some form of a fast attack ship. While he continued to serve, in December 1874, Cmdr. William Barker Cushing died in the presence of his wife in Washington, D.C.

He was 32 years old.

Uncle Sam

Jonathan Pessant

You know I used to think
I was the shit.
First in my family
enlisted: Army, active duty
standing in a field stabbing
a rubber dummy in uniform,
bloodshot-rust bayonet
into the heart of a tire.
See how it sucks
in the blade, as easily as a body.
Does blood really make grass grow?
It was barren in that field,
just the white-knuckled movements
of young hands.   Uncle Sam needs you
tattooed under our fingernails.

Your uncle was in the Reserves
my mom said one day after a few
flakes had coated the porch,
turning my footprints to slush.
I would be back to basic before they froze,
before I could think I could be wrong:
He never went to war; that doesn't count
These days   it's the seasons
my uncle remembers more than any war.

My uncle killed a man one night.
The man, already half frozen, I imagine
couldn't hear the steel plow
scraping snow from tar.
You see, my uncle was under contract.
Small town safe roads and freedom and the dream.
The man, curled up in a snowstorm,
the fresh blanket of 2am.

Slap-awake
and an arm in the air
and a leg in the snowbank
and the slam of brakes
and a slight skid
and the slow stare backwards,
the mirror red with brake lights.

I never killed a man though
I took an oath to do so.
In 2010, it was all the rage
to be called a combat vet.
Combat I guess means madness
I mean manliness.
It's all the same swerve
into the night   measure oneself
a length of dead-cold work
to be done for
Uncle Sam   Is there such a thing as knowing
you're a real man?

Combat Vet?　　　　　I imagine killing a man
gets you better service.

I can't imagine killing a man.
There's a sorrow of unexpectation,
of a weeping every January.
I imagine my uncle imagines
that I'm the lucky one
every time it begins
to snow.

When War Was in the Offing

Tom Nawrocki

The Location

To begin with, the Camp Pendleton United States Marine Corps military reservation is situated in the foothills of the San Margarita Mountains—roughly 123,000 square acres located directly between Los Angeles and San Diego. Its western boundary is the Pacific Ocean, and that part of Camp Pendleton with its pristine, sandy beaches and panoramic views of the golden, ocean sunsets could be a five-star resort. It is here that a complex of rifle ranges was developed for training Marines the skill of marksmanship. Each range was designed so boot camp recruits on the firing line turn their backs to the ocean—turn their backs to Vietnam in the offing—and face the scrub-filled foothills of the barren San Margarita range. They spend two weeks firing live ammunition for the first time at targets 200, 300, and 500 yards downrange. Here's where the story begins. Here's how it feels.

March, 1966

The days at the Camp Edson Marine Corps Boot Camp Rifle Range always began the same. You'd fall out on the road for chow well before sunrise at what the D.I.'s called "oh-dark-thirty," and a shroud of heavy fog would be floating in the breeze off the Pacific Ocean. The eight barracks buildings lined up at even intervals would be invisible except for the hushed halo of fluorescent light from inside. The mess hall down the hill toward the ocean and the flat, barren expanses of firing ranges further back behind the barracks up on the foothill apron would be totally obscured also.

64

But the fog itself had a real presence. You could actually see it drifting by you in the dark as you first found your squad leader and then your place in formation on the road in front of your barracks. This fog had a smell that sort of seeped into you while you were standing there zipping up your field jacket, buttoning all the loose buttons and checking your combat boots to make sure your trousers were bloused properly. You'd gradually feel dryness at the back of your throat from the saltiness of the sea moisture. That saltiness would get so strong at times that your eyes would begin to tear. The longer you stood there waiting in that dark, chilly air before marching to chow, the damper you felt. Sometimes you'd feel the concentration of water drops on your hands and cheeks and nose, and you might realize as the moats and strands of fog drifted by from right to left that you were actually standing in a cloud bank—that those drops of water forming on you were raindrops and you were becoming a part of that cloud and rain.

Eventually, Corporal Cherry would show up and bark, "Platoon. Ten-hut!" and if you weren't already at attention, you would snap-to with the sound of heels smacking together. You wouldn't see your drill instructor, only hear the sound of his commands coming at you from behind the platoon formation. His voice yelling at some recruit or calling for the fog guards to take position at the front and rear of the platoon, would sound somehow louder or nearer because it was bracketed by the fog. By the time you right-faced and got the command "fo-ward, haarsh," you were aware of the rest of the platoon. Even a thick mist couldn't obscure the back of Malloy's head, the same as always with the nubs of red hair and freckles on the back of his neck. You'll never forget that sight because you've had your eyes locked in position there during the endless hours of marching and drilling for these past four weeks of boot camp. Even with the early morning twilight now, you can still make out the mole on the left side of his hairline that still has a scab from the hack-job the last barber with fast clippers pulled on him. But the rest of the platoon in front and behind you marching downhill toward that invisible mess hall is a silhouette of starched covers and field jackets swaying in unison. You march along feeling the damp air on your face and hands while inside your field jacket you're working up a light sweat, and you can taste the salt suspended in that fog that has gathered over the Pacific and is a part of you now.

You can't hear the waves through the tromping of fifty-five pairs of combat boots marching in cadence, but you know those waves are out there (not more than half-a-mile away) just on the other side of Highway 101—just out of reach. Even if you hadn't seen the long, frothy columns of waves and the shimmering, reflected sunlight on the day your bus arrived, Cpl. Cherry would have described them to you, and also mentioned that it was just out of reach—out of reach like the surfboards, woody station wagons, beach parties, bonfires, and bikinied surfer girls who will be out there tonight after a long day on the beach. When the fog burns off in a few hours, they'll all be out there romping in the surf, but you won't. Cpl. Cherry has mentioned that on several occasions. You won't because you're in the United States Marine Corps now and stuck inside the Camp Pendleton military reservation wearing these drab green utility uniforms, sporting a buzz cut with white sidewalls in the era of shoulder-length hair, spending your days polishing boots and scouring out toilet bowls and drilling until you get heel contusions and holding out rifles at arm's length until your triceps feel white hot. So you won't be out on the beach today, "Lollygaggin' in the sun," as Cpl. Cherry puts it. Someone else will have to take your place—some surferboy with sun-bleached hair down to his shoulders and love beads and a running supply of reefer will take your spot in the sun and between that bikinied girl's legs because you made the God-awful mistake of enlisting in the Marine Corps and this spring day in March of 1966 you're not going anywhere until you finish bootcamp and sign your orders for the Nam. You know all that because Cpl. Cherry has said as much. And his word is the law around here.

Swinging Dicks

Lubrina Burton

Man: I was in the Army.
Woman: Me too!

Man: Well, I didn't sit in some cushy office; I drove them big ol' swinging dick Army trucks.
Woman: Me too.

Man: Well, every week I put on my battle rattle, and marched ass-deep into the woods.
Woman: Me too.

Man: Well, when I rucked, I humped over a hundred pounds of shit on my back.
Woman: Me too.

Man: Well, on marches, I was toting the M249 SAW, not that little toy, the M16.
Woman: Me too.

Man: Well, I fired grenade launchers and ATWs, the big guns that made men ruin their britches.
Woman: Me too.

Man: Well, I stayed in the field for weeks, through rain, snow, heat, just me and the boys.
Woman: Me too.

Man: Well, I slept in tents, washed my ass with baby wipes, and ate my chow from MREs.
Woman: Me too.

Man: Well, I worked with thousands of troops on a multi-national training mission overseas.
Woman: Me too.

Man: Well, once my sergeant took me off alone and dealt me some "corrective training."
Woman: Me too.

Man: Well, the Army taught me to be brave and how to die without fear for my country.
Woman: Me too.

Man: Well, I was still ready to kill me a few of them bastards in order to come home alive.
Woman: Me too.

Man: Well, because of all the shit I seen and done as a young soldier, I come back an old vet.
Woman: Me too.

Man: Well, what did you do in the Army anyhow that was so damn hard?
Woman:

Show Me What You Got

Ben Stewart

My first time on a commercial flight was July 25th, 2005 when I left home for the Navy. From civilian to military in the hour it takes to fly from Louisville to Chicago.

At O'Hare International, I followed the signs leading to the USO office where I joined a growing number of men and women awaiting their ride to Recruit Training Command Great Lakes. The room where we waited was good sized, but filled up quickly as flights arrived from all over the country and new recruits trickled in.

Whoever was in charge of this restless gaggle of humanity said to sit tight and wait for the shuttle, pointed us in the direction of the restrooms, and advised us not to wander. The few chairs and couch having already been filled with other people's butts, I found a place on the floor and sat looking through grainy photos on my flip phone, draining the battery and reminiscing on lazy days spent goofing off with my friends. Had I known at the time that I wouldn't get another chance at sleep for at least 24 hours, I would have taken a nap instead of strolling down memory lane.

A few hours of waiting in that crowded office and I'd had my fill of the USO. My phone now dead and small talk exhausted with those in range, I took to counting the stiches in the seam of my jeans when finally, the shuttle arrived and we were herded aboard. Thank God, I thought, let's get this show on the road.

We settled into our seats for the thirty-minute ride north. The driver had a few words for us, but I didn't hear what he said for one reason or another, then we got underway. We got on the highway and I realized how quiet it was, the bus's dark interior silent as the grave. Owing to a compulsion to fill any silence with the sound of my own voice, I turned to the guy next to me.

"Come here often?" I said to the guy, fishing for a laugh.

"Shut the fuck up," he hissed. "You're going to get us in trouble."

Tough crowd, I thought, but then realized *that* must have been what the driver was saying before we left, that he didn't want to hear a peep the whole ride or there'd be hell to pay.

After a quiet thirty-minute bus ride, we navigated a security checkpoint at the base and finally pulled up to a large brick building with a wide bank of glass doors. Imagine the main entrance to a large school or church and you've got the idea. Departing the bus, we were instructed by a yelling man in a crisp white uniform to enter the building with due haste and to stand on the blue tiles lining the main hallway.

The instant I crossed the threshold into that building, the English language I knew and spoke reasonably well was forever changed. Walls were now *bulkheads*, restrooms were *the head*, and the water fountain was a *scuttlebutt*. Call a wall anything but *bulkhead* and you'd get an earful. The main hallway was expansive and brightly lit, state flags hung overhead from poles fixed to the gray cinderblock walls and I found a spot to stand on the blue tiles as instructed. So far, so good.

Motivational anecdotes were painted about the large entryway; "Tough times don't last, tough people do" is one I specifically remember. My takeaway being "*This shitshow won't last, just keep your head down and you'll get through."*

Standing on my little square of blue tile, I waited to use the pay phone to call my home and tell my parents that I had arrived safely at my destination and would be in contact as soon as I was able. We were under very explicit instructions to say our name, confirm we'd arrived safely, and that we'd call in three weeks, then hang up without a single word of idle chit chat. If the instructor pacing the phone bank heard so much as an extra syllable than was necessary, maybe you asked about your girlfriend or pet goldfish or whateverthehell, the receiver was yanked from your grip and slammed into its cradle to terminate the call. Awaiting my turn, I saw this maneuver carried out more than once. Finally at the pay phone, I dialed home and my mother answered. I spit out a string of word vomit, "Hi it's me I made it I'll call you again in three weeks I love you goodbye."

She said "Okay, love you too." And I hurriedly jammed the receiver back in place and that was it. I was officially on my own.

After the phone call, I received the customary buzz-cut, processing quickly through and being shorn like live-stock, then it was back to the blue tiles and waiting my turn to get a uniform. Not the uniform I'd been imagin-ing, those iconic Crackerjack dress blue uniforms sailors are known for. Those, I soon learned, must be tailor-fitted and ordered for each recruit, which takes more time than was available for recruits fresh off the bus.

No, the uniform issued upon arrival at Great Lakes was utilitarian and designed for ease-of-use while pro-cessing through the Navy's initial medical screening to ensure you were fit for duty, i.e., you could handle intense physical exercise and you didn't show up to bootcamp with the Clap. The first uniform you got was called Smurfs and consisted of matching sweatsuits with NAVY written down the pantlegs and on the sweatshirt in reflective lettering.

The issuing of the Smurfs and the preceding buzz-cut, were the first visual representations of the military stripping away the concept of you as an individual and replacing it with the concept of being a unit. You were all dressed the same, sporting the same haircut; you would be treated the same no matter where you came from. Nobody was special.

While awaiting my turn, the enthusiastic drill instructors (called RDCs), divided us by gender; males lined up on one wall, erm, bulkhead, and females on the opposite. We were then filed into separate rooms and gathered our first articles of Navy issued clothing; Smurfs, white tube socks, white and gray New Balance's, white t-shirts, and several pairs of tighty-whiteys. Fashionable new clothes in hand, we found places behind square tables, empty but for a cardboard box and Sharpie for each man.

"I am Chief Soandso. Nothing I tell you to do is an option." A wide man with a shaved head and a stump of neck stalked up and down the rows of desks, thick hands clasped behind his back like he was barely restraining himself from throttling one of us. Maybe he was. "You will do what I say, when I say it, or you will be dealt with. Is that clear?" Muttered agreement from us new recruits. Not the response he was looking for.

"The answer to my question will either be *yes, Chief!*" he yelled. "Or *no, Chief!* Is that clear?"

"Yes, Chief!" I screamed, eyes forward and fixed on the middle distance.

Step by step, we were instructed to strip down and swap our civilian clothes for their Navy-issued replacements, old clothes put in the box to be mailed home to mommy. It's worth noting that we were forbidden to speak unless spoken to, our only focus was following instructions to drop trou and put on the Navy issued skivvies. What a sight we must have been, this room full of naked men donning tight and unfamiliar underwear, feet getting tangled in the elastic band and having to hop around, bare feet slapping linoleum, dangly bits flopping about, Chief Soandso yelling to hurry the fuck up.

"We're going to find out if you had a little too much fun before you got here," Chief said after we'd gotten dressed. "It's time to pee in a cup. Those of you who need to pee right now, line up on the starboard bulkhead. If you don't need to pee, port bulkhead."

Starboard? Port? A few recruits were savvy to what this guy was saying and hustled this way or that, following instructions with a *Yes Chief!* I was not one of those people. *Why, oh why*, I thought to myself, *didn't I study Navy terminology before joining the Navy?*

Thankfully not alone in my ignorance, there was confused hesitation as most of us didn't know what we were supposed to do, afraid of doing the wrong thing, instead doing nothing.

"Starboard means right!" he wailed, chopping a heavily muscled arm in that direction, then the opposite. "Port means left! Do it now!"

The room erupted into a scrambling mass of bodies as we found our places. I didn't have to pee, or at least didn't think I was ready to pee in a cup, which is a subtle but distinct difference. I went left and was herded into a large room with a water fountain (*scuttlebutt*, goddamnit).

Our instructions were clear. "You will walk around this room and take a drink from the scuttlebutt with every pass. Once you're ready to pee in a cup, raise your hand."

I don't know how long I was in that room or how many circuits I completed, but I finally raised my hand and was pulled from the shambling procession of recruits and ushered through a doorway across the hall. The room where you pee in a cup isn't private. Actually, privacy mostly doesn't exist in boot camp; modesty is stripped away with your civilian clothes the moment you drop your pants in a crowded room.

There were four urinals on a white cinderblock wall, like a public restroom, but absent barriers between. No privacy. A man paced behind you, barking orders and watching for anyone trying to cheat the urinalysis test. How you could manage that is anyone's guess, but the Navy taught me that procedures are in place for a reason and if there was someone watching your willy during the pee test, it's because someone has taped an IV bag of clean urine to their leg in an effort to cheat the urinalysis. You've got to be equal parts brave and stupid to give that a shot and I'd think just not doing drugs before reporting to bootcamp would be less effort than what it took to cheat, but I digress.

For the second time in the same night, I was instructed to pull my pants down, which, at the time, was a personal record. "Drop your pants and underwear to your ankles," the man said, still pacing and barking orders. The buttons on his white shirt were barely holding on, his uniform seemingly tailored to a slimmer past self. His voice was loud and echoed off the concrete walls. "Now step forward to the urinal, open your cup, and pee into the cup."

My teeth were floating, but I suddenly couldn't squeeze a single drop. Shit. Moments passed and those recruits at the urinal on either side of me finished and left, their full-bore streams filling the cup and thundering off porcelain. Mocking me.

After the second round of recruits came and went, I closed my eyes and focused on relaxing. Forcing yourself to relax so you can pee is rather unrelaxing, but finally a little pee dribbled out, working up to a steady flow. *Thank God*, I thought, but my triumph was quickly interrupted by the man proctoring this mess.

He yelled, "Number three!"

I opened my eyes and a giant number 3 painted above my urinal stared back. *Oh no*, I thought. *That's me.*

"Show me what you got!"

Exhausted and shaken by the preceding chaos, rattled by my newly discovered shy bladder, and being unknown to me how to turn off the tap now that it was going, a piece of advice my recruiter gave me floated up from the recesses of my consciousness: *Do exactly what you're told to do, when you're told to do it.* Then Chief Soandso's voice from earlier that night: *You will do what I say, when I say it, or you will be dealt with!* This threatening ticker tape scrolled across my mind's eye like an emergency broadcast, white letters in all caps on a red background, *Do what you're told.*

Without further thought or delay and with my pants bunched around my ankles, I shuffled around and showed him what I had. Which, of course, was my penis in one hand and a rapidly filling cup of piss in the other. Having tucked the hem of my t-shirt under my chin to avoid errant back-splashing sprinkles, he and everyone else were treated to an unobstructed view of what I had.

The look on his face as I presented myself to the room is something I'll never forget. His eyes went wide, his mouth froze in a surprised little O. He looked from me to little me then back up, his brain catching on to what I'd put on display. He stammered, then gathered himself. Inhaling through his nose so hard I felt the air rushing past, which was likely just coincidence and the breeze the result of standing bare-assed in a wide-open room, but still.

Once his lungs were full and he'd puffed up so much the buttons on his shirt threatened to pop off and shoot me dead, he yelled, "I don't want to see your dick!" Another breath, another burst of fury. "Show me what you got in the cup, you fucking clown!"

Embarrassed and muttering my apologies, I shuffled back around and held the cup, now full, over my shoulder for his review. He was giving me an earful back there, no longer shouting, but making himself heard to everyone around, *Goddamn moron* this, *Fucking dipshit* that. Seeing my full cup of pee and weary of my presence, he barked at me to pull my pants up and get the fuck out of his sight.

"Yes, sir," I said, fast-walking out of the room and avoiding eye contact.

Out of the handful of people who have yelled at me in anger, my mother still holds the title of most occurrences due to her being the primary disciplinarian of my parenting duo and my adolescence providing ample opportunities, but the guy I showed my wiener to on July 25th, 2005 holds the title for being the loudest and it's not even close. I left that room with a hard-earned cup of urine and ringing ears, also a valuable lesson in how loose phrasing can lead to unexpected outcomes.

My literal interpretation of *show me what you got* notwithstanding, you will find success in the military by simply doing what you're told and I never let this incident deter me from following instructions.

The Air Force

Sharnta Bullard

I left my parents' house at seventeen,
unsure of the world before me,
and signed up to be on America's greatest team.
I've always wanted to be a part of something bigger
    than myself.
I gave my all and conquered every challenge, big or small
traveled the world and pursued my education
met some of my best friends
I call family,
not related by blood,
but through sacrifice.
The ones beside me I trusted with my life.
These are the things that the Air Force gave me.

I never imagined how much my life would be changed by
    the military.
The Air Force afforded me many things, and for me, no
    other branch would ever do.
As I prepare to depart, retire,
I look back at all the things I was able to do,
and I am forever grateful to Air Force for changing my
    worldview
enhancing my life,
and instilling lessons forever imprinted on my heart.

Missing Man

W.J. Price

"Enough," I said, returning my ka-bar to its sheath. The blade seemed to hiss as it slid into the leather, as though resenting being forced away without first drawing blood.

Sergeant Wachsam looked up, not releasing the Imam's hair. "What?" he asked in confusion. Then, looking down at the sobbing holy man, "He's lying! He knows where the Mad Mortarman is!"

I sighed and turned away. "Yeah," I conceded. "Let him go." I took off my helmet to wipe away the sweat from the top of my head. Even in the shade of the Imam's living room, the Iraqi summer still set the air on fire. I glanced around while replacing my helmet and pocketing my salt-stained handkerchief. The Imam had the usual decorating taste as other upper-middle class Iraqis: lots of ostentatious, gold-covered tables and art, thick rugs on the concrete floor, and deeply-cushioned furniture. We'd drawn all the heavy curtains, of course, so there wasn't a breath of air to work against the heavy incense. I felt trapped in that low-ceilinged room; pressed-in by the eyes of family members smiling through framed photos and the occasional whimper of gagged, female voices drifting from the courtyard. In that moment, more than anything, I just wanted out.

Wachsam released the Imam with a grunt and a slap to the back of the bald cleric's head. "Fucking camel-humper," he growled, stomping his way through the heavy double-doors. I noticed Sergeant Laut lean in towards the Imam. "Remember, Hajj," he said in a gravelly voice, "this was just a bad dream."

Once outside, I stood apart from my squad, drawing a cigar from my shoulder pocket and trying to silent the nagging voice in my head, a voice babbling on about decency and morality. I ordered the gunners to release the Imam's wife and daughters. Cooke looked at me in disbelief, his 9mil already in-hand and ready. I just shook my head and repeated the unprecedented order. The women ran, weeping in terror, back into the house.

I could see my two team leaders arguing, though keeping their voices low. Wachsam was obviously angry, and probably frustrated. I wasn't, strangely. The Mad Mortarman, as we were calling the unknown insurgent that had, for the past week, been firing mortars at our base, needed to be caught; and he would be, eventually. Before, I would have torn our sector apart and done whatever needed doing to end the threat. Instead, as I stood under the shade of the Imam's palm trees, idly wondering how much water he needed to keep them alive, I just breathed in the burning, fresh air and let my eyes drift over the surrounding desert, beyond the walls of the house. I didn't feel the comfortable cold, calculating call to action that had always compelled me to some pretty inhuman acts. I felt…

Sergeant Laut joined me beside the massive water tank at the edge of the Imam's large courtyard. "How pissed is he?" I asked, nodding towards Wachsam.

"Very," my senior team leader replied. The older man retrieved his can of dip and took a pinch. "He doesn't get why you didn't finish it."

"*He* doesn't get…?" I looked pointedly at my friend.

Laut didn't look back at me, instead spitting on the Imam's spotless Mercedes sedan parked between us and the front door. "Yeah, *he* doesn't get it." My friend looked at me. "*I* think talkin' to that split-tail is starting to fuck with the job."

Had anyone else said that, at any other time… Instead, I just breathed deep the war-torn desert, trying to ignore the chlorine stench from the Imam's pool. "We don't have to be like this, you know," I said, mostly to myself.

Laut gave me a flat look and spit again, decorating the Imam's luxury car with his contempt. "I don't remember you sayin' that after the time with the Sheik's son. I don't remember you sayin' that when we went after those Iranians."

"Things change," I said, with a little uncertainty. "This place," I waved a gloved hand, stained by the country and my actions, towards the Imam's house. "It makes us as bad as any of them. We can be better."

"What was it you always said during training?" my friend asked me after another spit. "We come home together, alive and safe." He looked hard at me. "'Nothing else matters.'"

"We'll find the Mad Mortarman," I insisted. "We've got plenty of time. There's no need to take it to the next level."

Sergeant Laut may have had something to say in reply, but our conversation was interrupted by Sergeant Wachsam. The junior team leader was standing at the door of his gun-truck holding a handmike. "Boss!" he hollered. "Main's recalling everybody! There's been a mas-cas at the FOB!"

The base was in chaos. On approach, we could see a huge column of thick, acrid smoke rising into the hostile sky. Whole flocks of black medevac birds circled overhead, carrying casualties away. A herd of trucks stampeded back and forth from the landing area to what was left of the new food court. Alarms shrieked their tortured, ethereal wails. A chorus of shouted, often contradictory commands danced through the area.

My squad roared up, skidding to a halt in the ruined parking lot of the PX. We exploded out of our gun-trucks almost before the great metal beasts had come to a stop. None of the half-dozen trailers or pre-fab buildings were still standing, instead lying scattered about in mounds of burning debris. There was a mammoth hole in the side of the big PX, with most of the merchandise now scattered into air that carried the weight of burning gore and death. The whole area was a frenzied hive, frantically struggling to move debris and bring aid to the wounded, dying, and dead. Baleful smoke wrapped round our heads and slithered down our throats. The Mad Mortarman had hit us again.

My guys scrambled. Doc grabbed his aid bag and snatched up the three CLS-trained men. Their small group hauled ass towards the casualty staging area, where row upon row of moaning, screaming bodies covered in blood and pieces of their friends demanded attention from the harried medics. I ripped off my body armor and looked around. "Who's running this cluster?" I demanded in a voice likely heard all the way down in Kuwait.

"Who the fuck do you think?" First Sergeant roared right back. The elder NCO stood in the center of everything, trying to impose order by sheer will alone. "Who the fuck else would manage this fucking mess?" he demanded, sending a team of Intel Geeks into the smoking ruins of the PX. "Now get the fuck over to the food trailers and clear that fucking mess! There's still fucking soldiers trapped!"

We reached what had been our food court, but was now a burning, stinking, groaning mess of wall fragments, kitchen equipment, and chunks of what my brain insisted was burnt food. We began hauling on the mounds of plastic, concrete, wood, and glass, uncovering body after body, often piece by piece.

The sun continued its trip without a care for the tragedy below. Nobody even thought about a break. We worked until our uniforms dripped with sweat, and then we took off our shirts and kept working. Our brothers and sisters were trapped and we would find them. We didn't think about the one wounded person we found for every dozen bodies. There wasn't time for that. That was for later.

We only vaguely realized that night had come once some portable lights were brought in and turned on. The work continued as the shadows deepened, and darkness fell.

I was straining with a support beam so that Wachsam could free a weakly-struggling young commo-kid. Dorin, I think his name was, or Dor, maybe. My junior team leader had just managed to free the kid when Laut's voice came crashing over me. "Boss!" he called. "Over here!"

Having finally dropped that heavy-ass beam, I was bent over, taking deep breaths and arguing with my stupid heart, getting it back under control. I glanced over at my senior team leader. "In a minute," I said through deep breaths.

Laut looked at me then. Together, we'd seen a lot: friends killed, villages burned, women and children massacred. Laut and I had deployed together before this tour; we'd been through the shit. I'd seen him dead drunk and head-over-heels in love. I'd seen him in a murderous fury. I'd seen him bury his brother; another sergeant, like us. On that night, though, I saw something in him that stopped me cold. Something in his eyes… I moved over to stand next to my friend and look at what he'd found.

Buried in the rubble, a look of pain and horror forever locked on her crushed, unmoving face, was Bea Humphrey.

She was a counselor in the PTSD clinic they'd set up on our base. With so many guys developing symptoms and the media starting to report the lack of care, the Army had come up with the idea of having counselors in-theater, rather than waiting until we got back to start treatment. First Sergeant had noticed that I was getting a little twitchy, that my actions were getting… He ordered me to the clinic, ignoring my arguments that I was fine.

"You're not fine." That was what Bea had said to me. I'd made the point, right off, that there was no way for anyone to figure me out no matter how many hours we spent talking. "Actually," she had said through a bright smile, "I figured you out in about ten seconds."

"Isn't there some rule against you and me being… whatever, counseling. You know, since we… uh…"

"Since you tried to get me into bed?" she offered helpfully.

"I… that… it was only one lunch," I sputtered indignantly. She made me do that a lot, as I recall. "We never even…"

She had just leaned forward in her big Arab chair; I never found out where the hell she got it… damned thing looked like some medieval throne. She often waited until I was most flustered and then would just lean forward, her tiny hand on her tinier chin, and grin impishly. Her grin would make be blush, which was both humiliating and fucking infuriating. We spent two, almost three weeks like that, me trying to ignore things I didn't want to feel, and her making me feel things I didn't know I wanted to feel.

I went in several times, whenever missions allowed. Our conversations weren't counseling, I don't think. I would have therapists later, after I got out of the Army. Our time together didn't feel the same. We talked about things we liked and things we didn't, finding common tastes in bad movies and old books. Oddly, I remember Bea making me feel better about…

"You're not worthless, you know," she had told me on that last day, the last time we would speak.

I snorted, leaning back in the cushioned chair Bea kept in her office for the nuts like me. "If you only knew…" I could never finish that sentence. Not ever. Not with anyone.

"If I only knew the things you've done?" she finished. Bea wasn't even looking at me, instead fiddling on the laptop resting on her field desk against the far wall. "If I only knew the darkness of your sinful heart? If I only knew the terrible acts you've committed for king and country?" She glanced across her narrow shoulder at me, brilliant eyes clear against a golden halo of short hair. "For the millionth time. I. Don't. Care."

"Well someone should," I muttered, glancing down and away.

Bea finished what she was doing on her computer and stood, facing me. "Stand up," she commanded. Music began filling her small office-trailer, a soft horn playing a simple tune.

"What?" My mouth objected while my body hurried

to obey. I hated that; it really annoyed the crap out of me. That damned Jedi mind trick or whatever she did. Who the hell was the itsy-bitsy girl that thought she could order *me* around? I think, looking back, I was most annoyed with how *quick* I was to obey.

"Come here," she instructed, holding her arms wide as the soft music reached out and closed the distance between us.

"I…what?" Again, my mouth objected while my body obeyed. God, that was annoying.

She pulled me in close, so that I was pressing against her. I started to say something, I don't remember what, but suddenly realized that we were both only wearing PTs, the cotton t-shirts and shorts offering little barrier to the warmth that seemed to flow out of her and into me. I was shivering, despite the heat of the Iraqi morning, and increasingly confused by a swirling feeling in my chest, a feeling I wasn't sure was unwelcome.

"Pay attention," she said, tapping a finger on my thick forehead.

"Hey," I objected. "You're the one getting all… what the hell are we doing again?"

"This is a lesson for you," Bea said, soft eyes grabbing hold of mine.

"In… dancing?" Something about those eyes, I remember. So blue and…

"In being human," she replied. She put her head against my heart, as though listening to the frenzied thumping. Her scent filled my mind, a scent of warm mornings and laughter. "I don't know what you've done that makes you hate yourself," she said so gently I felt it more than heard. "I don't know what you're capable of, out there. I don't want to know because it doesn't matter." She leaned back, just a bit, and locked my soul within two gentle, flawless blue pools. "*This* is what matters," she said. "You choose, every minute of every day, what kind of person you are. You can be a monster, or you can be a person. And, if you are person, then you deserve to be loved."

I don't know, I didn't then and I don't now, where those drops of salty water came from that fell on her delicate, caring face. I don't know how long we stayed like that, swaying to some melody only we heard. I like to think that, in some perfect universe, we were like that forever, or at least for more than that one moment.

The next time I saw her, Bea had been crushed under a building.

There was no talk of vengeance. There were no soaring speeches, swearing justice would be served over the broken body of the woman…over the body of a good woman who deserved better. There was no broken, desperate sobbing or pleading with an uncaring God to undo what had been done. There was nothing; the welcome, comforting emptiness returned to me as though I had never tried forcing it away. It banished the unhelpful, noxious emotions that had threatened to change me from what I needed to be. The soul that had started to sprout was thankfully uprooted and tossed aside.

I do not remember how long I stood there, over the body. A while, I suppose. When I called the medics over, they appeared immediately, as though waiting. I did not linger to watch them cover the body or transport it to the field morgue; I had work.

When undertaking certain missions, we found it

useful to have alternate equipment. We did not wear the grey ACUs that American soldiers were using at the time; instead, we wore dark green BDUs often used by the various paramilitary companies operating in and around Iraq. The uniforms were sterile: they had no name tapes, no rank insignia, no unit patches, and no American flags weighing down our right shoulders. We secured our M4s and M240s with 1st squad, instead pulling out the various AK-47s, pistols, and shotguns we had confiscated over the past six months. We also brought our knives. Instead of the easily-recognizable American body armor we normally wore, we pulled out the smaller, black ASR vests I had purchased with private funds for certain missions.

We waited until full-dark of that moonless night. The Imam owned a large, walled house near the village of Al-Dul. By the time we got there, rolling gently with blackout lights, even the latest of the local night owls had gone to bed, as we had planned. Sergeant Laut, using his PVS-14, identified two guards at the gate and another inside the wall, roving. No doubt after our visit that afternoon, the Imam felt the need for security. I left the drivers with the gun-trucks, keeping our egress ready and secure. I brought no attachments that night; I needed no interpreter and Doc had work at the base. I took the gunners and team leaders and we infiltrated through the back, silently going up and over the wall. Sergeant Wachsam slid up behind the roving guard and slit his throat. I realized, at the time, that I would have to talk to him about that; yes, cutting the throat prevented a scream, but it also exposed your torso to a counter-attack and caused a big burst of blood. Once the roving guard was neutralized, Sergeant Laut and I moved around to the gate. I grabbed the target on the left while Laut went right; I put mine down with two quick, twisting blade-strikes into the target's back, deflating the lungs. Sergeant Laut covered his target's mouth and reached around, stabbing high into the chest. Again, I considered future training for the squad; they were too dependent on those big, fast kills. Admittedly, opening the throat or rupturing the aorta would cause a quicker death than striking the lungs, but it was also messier and carried the potential for counter-attack. Better by far to hit low and let the target die silently, if slower, without risking yourself. But that was for later.

With all three security elements neutralized, we moved towards the house. I had half the squad stack on the main door while Sergeant Laut took the rest to the other ingress points, making sure nobody could escape. It was over almost before it began. Wachsam kicked the door and I led the squad inside. Laut led his section in through the back door. There were two men inside that were not the target, so they were eliminated. The wife and two daughters were secured and taken outside, to kneel by the gate. I had the gunners pull security on the women, removing them from the house; the witnesses would be eliminated later, once we were ready to leave. This left only myself and Sergeants Laut and Wachsam.

We dragged the Imam into his living room again, binding him to the same chair as before; it still stank of his urine from the afternoon's visit. When he tried to speak or curse or just scream despite the gags, one of us beat him until he remained silent. I knelt in front of the Imam and drew out my knife. "I know you can speak English," I said in a voice lacking any humanity. "If you say one word in Arabic, you will be harmed." The problem with having made a threat earlier and not following through, is that the subject will doubt your willingness to proceed. Therefore, an opening must be established wherein you can demonstrate the absolute certainty of your actions. The Imam started saying something in Arabic, as I suspected he would, so I cut off his smallest finger. My credibility with the subject was thus restored.

Sergeant Wachsam held a gloved hand over the Imam's mouth until the cleric quit screaming. Some people are surprised how quickly most subjects will give up on screaming, how quickly they give in to their despair; I, however, had experience with interrogation. Once I again had the Imam's unblinking attention, I placed the tip of my knife against his groin. "Where is he?" I asked calmly. "Where is the Mad Mortarman?" In response, the cleric started to claim he did not know, so I pushed the blade in about a quarter inch, not far enough to kill or risk quick death, but enough to cause horror, pain, and humiliation. Again, Sergeant Wachsam held the Imam's mouth until he quit screaming. I repeated the question, "Where is he?" and, when the cleric did anything besides give an immediate and honest answer, I twisted the knife about ten degrees.

Then, the Imam talked.

Less than an hour before dawn, our gun-trucks pulled up to the village of Antiquam, a Shiite village. The air was cool, as cool as it could get during the Iraqi summer, and the dim, pre-dawn light painted the sky like fresh, venous blood. Several of the villagers, the strong men of the community, were waiting for us in the marketplace. We did not need to call them, instead driving with our white lights on during the approach. Their lookouts spotted us, as we knew they would, and alerted the local militia. Thus, a dozen or so burley-looking Shiites carrying an assortment of AK-47s and even a PKC were ready to welcome us.

We came to a stop and, without a word to the assembled thugs, I turned and went to the cargo hatch. My two team leaders did the same, each retrieving a large bundle from their own trucks. I dragged the man that had been tied and gagged in the back of my gun-truck towards the waiting Shiites, tossing him like so much garbage at their feet. My team leaders did the same with the other two.

The leader of the Shiites reached down and pulled off the hood from the man I had tossed at his feet. The target blinked up through blood-soaked, swollen eyes. His beard had been shorn off with jagged cuts from very sharp knives. Each of the men was naked, their clothes having been shorn away in a similar manner. Their bodies were a kaleidoscope of cuts, bruises, open wounds, and obviously-broken bones. Each of them sobbed softly and tried to curl into themselves.

The Shiite leader looked up at me and nodded. He recognized them. The three that had been attacking us as the Mad Mortarman were, in fact, Saddam loyalists, former Republican Guard that had spent years terrorizing this very village. Only the American presence had kept them safe from Shiite retaliation. I nodded back and my team leaders and I returned to our gun-trucks, leaving the final disposal to the locals. Eventually, three headless bodies would be fished out of the Tigris, but that was later.

We were silent as we formed up a few days later in what had been the PX parking lot. Deep shadows stretched their claws over us, strong and cool but retreating quickly from the rising sun. The casualty collection point had been cleaned as best they could once the last of the bodies were taken away. Still, you did not have to look hard to spot the stains of suffering and death that would forever paint the place. There were no birds, there rarely were in the desert; the sky was empty, as empty as those of us still on the ground. A line of battlefield crosses were set up against the ruins of the PX, one for each of the fallen.

I let my body go through the mechanics of the memorial ceremony. I went to attention when they posted the colors. I bowed my head when the Chaplain gave the invocation and Scripture reading. I heard none of it, nor had any real sense of presence. Finally, the clerk came forward and read the Last Roll Call. Name after name filled the air; the list of the fallen. Finally:

"Beatrice Humphrey, 2nd Lieutenant."

I stood there, silent and unfeeling. Trying to understand… I did not care about the anger that some felt, or the sorrow. I only wanted to get on with the work. I had tried caring but found the price too high. I took a deep breath as the clerk read the name again.

"Beatrice Humphrey, 2nd Lieutenant."

Perhaps it would have been better had we never met. She would still be dead, of course, but then I would not have been burdened by her death. Then again, had I never met her, I would never have tried to fight a war with compassion. Had I taken the job seriously, without unnecessary burdens, I could have stopped the Mad Mortarman before she died; then she… The clerk read her name one last time.

"Beatrice Humphrey, 2nd Lieutenant."

I let my eyes drift to the front of the formation, where they had placed her battlefield cross. Her rifle was propped up, barrel-down, between a pair of her boots. They had to be her boots, as small as they were. Her undersized helmet rested atop the butt of her rifle, and her dog tags tangled down, against the M16. The bugler began playing taps, a soft piece of music that drifted in, amongst the formation. She would have lived and helped a great many broken soldiers, perhaps even…

The bugler finished, and the formation was dismissed. Some people went forward to cry over the small memorial or say some quiet farewell. Many eyes went to me, expectantly.

The missing man table would be set for dinner. The chow tent was silent as we filed in that evening. The front of the dining area, normally an uneven mass of dessert trays, trash cans, and unused guest tables, had been cleared. My squad and I filed in, joining those who were also coming to pay their respects at the final meal. A place had been set aside for me and my men, very near the front, seconded only by the assembled, high-ranking officers. Only the grinding whir of the overhead fans disturbed the heavy pressure of the tent, and even they seemed subdued by the ceremony at hand. At the front, separated from the rows of tables for the living, was a small place for the fallen.

An honor guard marched in, slowly, carefully, silently. They marched to the front, where the small table round stood alone, round to show our devotion and equality in death. A clean white tablecloth was draped over, white for the purity of her service and sacrifice. Five places were set, one for each branch and another for the civilians. A single rose was placed in the center, red for her blood. A yellow ribbon was tied around the rose, the remembrance. A slice of lemon was placed on each plate, the bitterness of our loss. A pinch of salt fell over the lemon, my tears. A single, white candle was set in the center of the table, a hope for reunion. Each plate received a glass, inverted since we could not share a toast. Five empty chairs rested around the table, places of honor for absent… A bell was struck three times, accompanied with the word: "Remember."

We ate in silence, I did not trust my voice, just then, and others had respect enough to honor those missing. Many glances were cast towards me, and towards the missing man table. I did not acknowledge the glances, or the table, I did not know that I could. I gave only the most perfunctory replies to the officers and senior NCOs who mumbled some sympathy at me. I sat with my men, and then I left, wanting nothing more than out.

It was full dark when I returned to the temporary memorial at the PX, now still and empty, save for the battlefield crosses. I had wandered around the base until then, not really going anywhere, not really thinking anything; I had no job, just then, no purpose. I was a little surprised that my wanderings had brought me to the PX; it was probably habit, from having walked there so many times before. There were a few others, gathered around this cross or that, some weeping, some not. I did not bother with them, I walked up to the only cross that really mattered, in that moment. I stood there, looking at the small memorial, not feeling…

I thought I could hear whispers behind me. *There goes the fool who felt*, they said. *There goes the fool who forgot. There goes the fool who failed.* And they were right; all the whispers were right. I was a fool. I had forgotten and I had failed. I had made the mistake of feeling and someone better than me paid the price. I stood there, at that small memorial, for a long time. I let the whispers go on, reminding me…

When Dad Went to War

Mark Madigan

There were occasional
handwritten letters—
scribbled in familiar

black felt tip—
but they didn't come
often enough.

Each day, after school,
we'd race to the mailbox
clinging to the house

but the most frequent sound
other than the rough
shuffling of paper—

the junk mail, bills,
and letters for Mom—
was the black metal mailbox

lid slapping shut,
the sound of our emptiness
heard in a clink.

Snapshots

Mark Madigan

I wasn't supposed
to see those glossy
black-and-white photos
falling from Dad's
faded duffel bag

but the first one I touched
showed Dad with an unknown
Army friend
standing by the body
of a dead VC,
smiling.

In the next, three soldiers,
out on patrol,
were walking single file
with rifles held tight
against their chests
as if they expected
something to happen.

With Dad in the lead,
the team headed down
a narrow dirt path
toward the camera,
a close, clear shot.

I couldn't help
but imagine crosshairs
superimposed
atop the photo.

Then I had a vision
of two unknown
Asian men standing beside
my father's body.

I was startled as Dad
scooped the photos up,
barking they weren't
for children to see.

I said *Sorry,*
but he couldn't take back
what I'd seen,
and I never told him
about the dark things
I still imagine.

Flashback

Mark Madigan

We never saw signs
the war had changed him,
until that Saturday,
after running errands—

as the car dipped down
beneath a beltway
underpass—

when someone in a road crew
working above us
let the sharp blast
of a jackhammer go.

The moment Dad heard
that sharp rat-a-tat
he yelled *Duck!*

and stirring up a cloud
of gravel and dust

zoomed the car up
the shoulder of the road

before it lurched
up over a curb

and skidded to a stop
in a church parking lot.

As Dad turned around—
checking on the kids who'd
tumbled about—
I saw his hands

shivering nearly
as bad as Gran's,
even as he turned back
cursing as he gripped
the steering wheel again.

House of Delight

Peter Geier

I HAD SEEN BORDELLOS IN MOVIES and read about them in literature. I had seen the "beaver displays" in the area around Place Pigalle in Paris, similar things in San Francisco and New Orleans, and been solicited on the street. But I never had been inside such a place or paid to have sex: I was curious about these things but never actually expected to do either.

While I served with the Army in Turkey in the 1980s, Bill, a guy in my section, invited me to go with him to a "whorehouse". This place was in Çorum, a rural provincial capital south of our duty station in Sinop. Bill had heard about it from Navy senior chief petty officers, which in itself probably said all one needed to know. Bill liked to talk about going to such places when he was stationed in Berlin and wanted to check this one out.

I supposed that Bill had asked me because he took me for a fellow sophisticate: I would not let on to being anything less. This was bound to be an interesting adventure. I had read a lot about Turkey, spoke the language a little and followed current events. My experience of provincial Turkey was that while people were friendly and generous, the mainstream culture was religiously observant, socially conservative, and strictly segregated by sex despite fifty years of state secularism. My guess was that our proposed destination would be a far cry from any red-plush-and-gilt-mirror experience that Bill presumably had in Berlin. If there were such a thing as a true Turkish house of delight where we were, I was curious to see it.

Getting there would not be easy. Çorum is 100 miles due south of Sinop as the crow flies, though it was more than twice that far by the existing basic primary and secondary roads. We could take a bus or cab, go down on a Saturday and come back on Sunday. But then there was the problem of finding the place, and neither Bill nor I spoke Turkish enough to get out of a jam. However, Cem, a Turkish soldier who worked with us, knew about the place. He said that a taxi-driver friend could take us. Cem, who was married, was all for the trip. He said we could do it in a day.

As the excited talk and trip planning progressed, I sensed that this would be less a sexual escapade than something like Jack Kerouac and friends piling into a car and driving to a border town in Mexico, beat and gone less the marijuana. I am always up for a road trip. Our route would take us into the heart of Hittite country. There were historic places to see along the way; all of it would be new ground to me. In fact, my expectations regarding the prospective house of delight were so low that I invited Terry, a friend and married Army colleague whose wife was a friend also stationed in Sinop. Terry was a soft-spoken, educated, level-headed Southerner. He was fluent in Turkish that he had taught himself and interested in the country and its culture. I thought he might share my curiosity about seeing something like this. I brought it up while having a meal with him and his wife. They goggled and gaped at me as though I were kidding or had lost my mind. I admitted that it might sound crazy to ask, but I was sure that this trip would be "one for the books". I seriously doubted that there would be any sex and would not have asked Terry otherwise. He declined; the three of us remained friends. Bill got another colleague, Mike, to go instead. Mike was a cautious, low-key married guy who I think went mainly to see what would happen because Bill and I were going.

Five of us left early one Saturday morning on a raw day in early spring: Bill, Mike, and I, along with Cem and Ferhat, the taxi driver, who may have been Cem's cousin. Ferhat and Cem rode in front and the three Americans in back of the Renault sedan taxi. Our route took us from Sinop along the coast road toward Samsun, south through the mountains to Amasya and then west to Çorum, easily 200 miles. The empty landscapes could have been the time of year: earthen towns and villages with long stretches between, a plane tree here and there and rock outcroppings along roads cut through the hills. We stopped for tea at a çayevi, which then in Turkey were landmarks for truckers and travelers the way diners used to be in the U.S. The people I saw looked less like the Westernized Turks I had come to know and more like people I had seen in photos of rural Central Asia and Iran. We made a brief restroom-and-snack stop in Amasya by late morning. This historic town is famous for its location at a strategic mountain pass. There was an old mosque on the main road in an institutional setting with a shrine, a fountain, and a museum that may have been open, but our Turkish companions pressed us to move on. From the car windows I saw no indication of anything Hittite the whole way to Çorum, certainly nothing I had hoped to see as dramatic as ruins such as the Lion Gate of Boğazköy. These are well beyond Çorum to the south and probably not visible from the road anyway.

The Çorum I saw was a charmless Turkish provincial seat with Atatürk images and his sayings flying in white letters on red banners typical of that time when the country was under military rule. I could squint and pretend I was in early Soviet Russia. It was far from Europe; the town did not appear to be geared even toward Turkish visitors. Ferhat drove us through Çorum to a place on the outskirts where trucks were parked on a mud lot near a large institutional structure. The trucks were colorfully ornamented and the drivers, older men in skull caps with weathered faces, made Cem and Ferhat seem almost American. Bill and I shared a raised eyebrow.

And our destination house of delight turned out to be a house of correction: The large institutional structure was a women's prison that opened its doors to paying customers on weekends. There were guards at the prison gates, but I recall no one checking us for weapons or contraband. Cem may have vouched for us; we were all in civilian clothes and I do not recall anyone asking even to see our identification. A prison matron escorted us inside to an area like a school gymnasium. On the way, Bill pulled out a bottle of Jack Daniels "to get the party going." Our escort angrily rebuked him that alcohol was forbidden and to put the bottle away.

What we saw in the gym was beyond anything we had imagined. There were as many as a dozen booths in two rows, six booths per side, each of which formed a seven-foot cube. The booths were hung with sheets, open on top and covered in front with a sheet curtain; each had within a cot and a large plastic basin. Skinny, pale, forlorn-looking young women in hand-me-down polyester slips stood in front of the curtains. "Mavi mavi," a popular Turkish Arabesk song roughly equivalent to U.S. honky-tonk country music, played on a boom box. The madam was a brassy young Turkish copper-blonde in a leopard skin one-piece bathing suit, Tarzan-style with a single shoulder strap, and gold pumps. Tarzana engaged in ribald verbal exchanges with the men in line as she talked up her "girls" and took the men's cash. Men would choose one of the women in slips and go with her briefly behind her curtain. The going rate was not more than the equivalent of a dollar or two. For the women in slips, the routine appeared to be wham-bam, rinse and repeat.

Cem and Ferhat rushed ahead to get in line. Bill's usual cool, seen-it-all, done-it-all look for once disappeared. "This is un-fucking-believable, man," he said. "I've never seen anything like this in my whole life, and I've seen some pretty wild-and-crazy shit." Mike stared at the scene with his mouth open. Middle-aged Turks in civilian clothes standing nearby made saccharine smiles at us and offered gestures with their hands toward the women inmates as though we were a bit dim and needed prompting. The plain face of an inmate with whom I made brief eye contact has stayed with me like a photograph. After several turns, Cem and Ferhat asked why we had not joined the fun. Cem was disappointed when I said that this was not what we had in mind. He explained that the prison does not feed inmates: This is how they get money to buy food, bad women whose families have abandoned them. "Bad women," he said with a dismissive gesture which said they are nothing, nobodies, that they don't count.

I heard later that Turkish prison inmates' families either moved or moved a family member near a prison or paid a local to ensure that their incarcerated relative got fed. Women whose criminal actions had "disgraced" their families apparently did not necessarily receive the same support. To put the local food and money in perspective, I recall paying for a full restaurant meal with raki in

Sinop in a small private room with several people that totaled less than the Turkish lira equivalent of 30 U.S. dollars. I took home roughly $800 a month as a buck sergeant and the Army fed, clothed, and housed us.

Bill and I considered paying to go behind a curtain not for sex but to help the women get their food money; maybe we could slip them something extra. But this would just as likely confuse them or even get them in trouble: We could not adequately explain our intentions and did not want to be any further involved in this pathetic travesty. Bill gave Cem and Ferhat what to us was like Monopoly money when they ran out. "This is one sad fucking commentary, man," he said, shaking his head. "And about as sexy as beating off with a mop."

Waiting for Cem and Ferhat to finish, we kidded Mike about taking on Tarzana, the brassy lass in leopard skin with the fat fistful of lira. She appeared to be an official not on offer but we did not ask. We were relieved to get escorted out after Cem and Ferhat had their fill. Ferhat was too sexed-out to drive so he and Cem slept in the back seat while I took over behind the wheel with Bill riding shotgun. Since we had not had lunch, the plan was to find a good place on the road for dinner.

Until then, I only knew Bill socially because we worked different shifts. I often have found that sharing an experience like the one we just had tends to lead people to drop pretenses and speak from the heart about things on their mind awakened by what they went through together as though to a close friend or family member. Bill and I spent the rest of the afternoon drive talking in this vein. We had no trouble getting to Amasya, but I was not used to driving in Turkey and at first found Turkish drivers hard to predict. Things got more interesting after dark because for some reason oncoming vehicles flipped on their high beams just before passing us: I got the hang of it and tried to beat them to it. It had rained while we were inside the prison. The most incredible part of our afternoon drive, maybe the whole day, was not only that we saw a perfect double rainbow over the road ahead, but also drove under it.

We did not see a place to eat until Amasya came into view just before dark. High on a cliff to the right I spotted a building with large windows that looked as though it could be a restaurant. "That's where we're going to eat," I said. But the place turned out to be closed, condemned because an earthquake had loosened its moorings, making it unsafe. Everything else in town was closed. Cem spoke with local police and soldiers at the town headquarters. The country was under martial law at the time. An officer invited us as soldiers to be his guests and got someone to open a restaurant for the occasion. Other men showed up. We had a huge meal and drank raki.

I drove us home afterward talking with Bill and playing the headlight game with sporadic oncoming traffic while the other three slept in back.

Mail Call

LeeAnn Meadows

*After World War II letters
from Ralph to Marilyn
November 1944–February 1945*

Mail call today, my sweet,
and still no word from you—
It is hot on the ship.

At mail call, everything stops
until all the letters
are read and reread.
I always hope for mail,
a guy needs to hear about home.
Took a nap yesterday,
send pictures of the girls.

No mail for two weeks.
I write anyway to build character
so you will have to listen.
You embarrass me—
One loses face with no mail.

Mail call today,
your first letter arrived,
so good to hear from you.
Kiss the girls for me,
send Emery boards
and Kenyan blades;
there are plenty of cigarettes
on the ship.

Mail is erratic.
I read history backward
in three letters from you,
a welcome gift from the sky,
something to fight for.
In your letter of the fifteenth
your tone much better,
keep up the good word.

Mail call today—
so hard to write when
I don't have a letter from you.
Time chips away,
mail closes in ten minutes,
I will get this letter out.
I miss you,
your Ralph

"Detective Chicken Reporting for Duty, Sir!"

Donald Delver

AS A YOUNG SERGEANT, AND THE newest instructor at the 15th MP Brigade Academy in Mannheim, Germany, I worked hard to prove myself to my superiors and to the experienced Military Policemen and Police-women who attended our training course for MP Investigators. For example, I helped some of the other instructors who had volunteered to build two new classrooms on the top floor of the old German Army barracks that was our home. They kidded me good-naturedly about my lack of carpentry skills, but I matched them hour for hour and beer for beer. Pabst Blue Ribbon was the beer of choice for thirsty MPs working overtime on Army pay and we filled the hollow spaces in the walls of the new classrooms with a shitload of empty PBR cans.

I took my work seriously and studied hard to prepare for my classes. I was the primary instructor for Interviews and Interrogations, and I also prepared to back up the instructors for Drug Investigations and Elements of Offenses. As a younger sergeant, I probably took myself and my work a bit too seriously, fearful of being caught in a mistake by an older, more experienced student. Sure enough, during my first Elements of Offenses class, the room was filled with many experienced MPs, a few of them looking to have a bit of fun with the young instructor up on the platform. This class was the genesis of Detective Chicken, though I didn't know it at the time.

During the lesson, I had gotten as far as Article 125 of the Uniform Code of Military Justice without incident.

That article covered the elements of sodomy, and after a brief review of the elements, one of the older sergeants, a huge smirk on his face, asked me to give an example of sodomy. I responded, "If you were to make love to a chicken, that would be considered sodomy under Article 125." That sent most of the class into hysterics, and after a minute or two, I continued with the lesson and forgot about the incident until a few weeks later at the graduation ceremony.

It was the custom for graduating classes to give presents to their instructors. Sometimes the presents were thoughtful and sometimes they were humorous. While the instructors stood on the raised teaching platform, one by one, a student would come forward and present an instructor with a present. This class presented me with a good-sized hen with vivid reddish-brown feathers, a bright red comb and alert eyes that looked back at me from the cardboard box the student handed me. I took the box in the spirit of good fun, thinking to myself, "What the fuck am I going to do with this?" There was laughter and goodwill all around, and after a brief formal ceremony, graduation was over, and the students went downstairs to pack.

After the students left, I took the cardboard box with the large hen inside downstairs with the thought of finding a farmer who might take it off my hands. Before I could leave the building, however, a clerk saw me and asked, "Sergeant Delver, what are you doing with that chicken?" Still chuckling to myself from the joke the class had played on me, I decided to tell the clerk a tall tale and have a bit of fun with him. I asked him if he could keep a secret, and when he said yes, I moved closer and spoke softly, as if taking him into my confidence.

"You know that the military uses dogs to sniff for drugs at airbases and other military installations, right?" He knew. "Well," I said, looking around as if to make sure we were not observed, "what you may not know, is that the nasal membranes of dogs, while highly sensitive, cannot reliably pick up the scents of certain hallucinogenic drugs. We at the Academy have been working to find an animal that can detect what a dog may miss, and the nasal membranes of chickens are quite sensitive to a variety of hallucinogenics." With that, I put my finger to my lips, to remind him to keep silent about this extraordinary information I had just made up, and I went about finding a farmer who would take the chicken off my hands.

I thought nothing more about the chicken or the encounter with the clerk until a few months later when I was on leave in Garmisch Partenkirchen, a ski resort in the Bavarian Alps. I ran into someone I knew, a soldier who worked in the motor pool, and I asked him how he was enjoying his time off. We chatted for a bit, and then I asked him, casually, how the drug scene in the Garmisch area looked. "Not great," he replied. "The Army is using chickens now to sniff for drugs." I had to bite my lip to keep from laughing in his face.

"Really," I said. "You'd better be extra careful from now on, then." I said goodbye and walked back to my hotel, chuckling to myself the whole way. My job from then on was a lot more fun because I no longer took myself quite so seriously.

Surviving Christmas

Bruce Sydow

A young Marine endures combat
with a flinching hyper-vigilance,
clenched-jawed and courageous.

He relives it many years later
with the debilitating sounds
of fireworks and party poppers.

The old Marine's life has become
incapacitated in a startling web
of paralyzing noises.

But a sunrise applauds
the hope of possibilities
as dawn breaks over the Cascades.

I knock on the grizzled veteran's door,
because no one with PTSD
should be alone on Christmas.

Grimacing at the intrusion,
his heart pounds like a drummer boy
as he peers into the peephole.

The door opens a crack, then wider.
I smile and whisper, "Good morning,"
and hold up a plate of warm cookies.

"Look, red and green sprinkles.
My wife made them special,
just for you, friend."

A plaintive note constricts his voice
and bursts into a paroxysm of sobs
as he struggles to say, "Thank you."

The sun, in full-throated glory,
suddenly bathes the foyer
and engulfs him in renewal.

And in that cleansing moment
I know there is hope,
for both of us.

Contributors

Sharnta Bullard started writing poetry at a very young age and had her first poem published at age eleven. Sharnta has poems published in several anthologies. Additionally, she is a United States Air Force retiree and has a master's degree in Human Service Counseling. She is an advocate of helping people find their why and voice.

Lubrina Burton is a veteran of the United States Army. Her experience as a young enlisted soldier inspires much of her work. After the Army, she returned home to Kentucky where she studied creative writing at the Carnegie Center's Author Academy in Lexington. Her short pieces are featured in *That Southern Thing, Trouble, Curious Stuff, Twists and Turns, From Pen to Page II,* and *Alice Says Go Fuck Yourself.*

Called the "blue collar writer" by peers at the University of Central Florida, **Bill Cushing** moved to California after earning his MFA from Goddard College. Retired after over 25 years of teaching, he has four published volumes of poetry and his first collection of short stories, *The Commies Come to Waterton*, was released last summer. Bill is revising a memoir focused on his days serving in the Navy and working on commercial ships.

Donald Delver spent nine years on active duty as an Army MP during and after the Vietnam War Era. He currently resides in Oregon where he teaches English to recent immigrants as a volunteer. His short fiction has appeared in the anthology *Metamorphosis: A Collection of Stories*, published by Propertius Press, and in the March 30, 2023 edition of *Flash Fiction Magazine*.

Noelle Dennard is a former "Army brat" who has been writing poetry since she was small. Growing up, she was always inspired by the adventure of moving, the various landscapes, and her family that always proved home is where the heart is.

Michelle DeRose teaches creative writing and Irish, African-American, and world literature at Aquinas College in Grand Rapids, Michigan. She is the daughter of a former service member: her father served as the Company Commander of the 37th Medical from June 1967-68. Some of her poetry about her family's experiences during and since his deployment can be found in *As You Were: The Military Review* and in *Bear River Review*.

Mitzi Dorton is author of the book, *Chief Corn Tassel* (Finishing Line Press). Her poetry is in *Rattle*, Willowdown Books, *Women Speak*, Women of Appalachia Project, and her work is forthcoming in *Poetry South* and Arachne Press. A multi-genre writer, she has also been published in many literary journals and several anthologies. She is the daughter of a World War II, Hell on Wheels, second division Army veteran, a tank driver in the Battle of the Bulge.

Aliza Dube is a military spouse and a student of the University of Southern Maine's MFA program in creative writing. She is the author of *The Newly Tattooed's Guide to Aftercare* through Running Wild Press.

Peter Geier served as a U.S. Army linguist from 1981-85. He has worked as an award-winning journalist, reporting and writing features for local, national, and international publications. He has published two pieces of short fiction and one of creative nonfiction. He reviews films for http://moompitchers.blogspot.com and is completing a collection of stories set in New York.

Mary Gutierrez is a Hispanic female whose parents migrated to the U.S. in search of the American dream. Her grandfather, father, brother, and brother's son served in the U.S. military. Mary has been published in numerous local and national poetry zines and magazines. She is the author of *Naked in the Rain* and is part of the anthologies *Word in a Web* and *Voices of the Earth: The Future of Our Planet III.*

Kathleen Isaac-Luke was born in Baton Rouge Louisiana and has a Master's Degree in Nursing from Texas Woman's University. From 1966 to 1968, she served in the U.S. Army Nurse Corps in Bad Cannstatt, Germany. She currently lives in Sonora, California and was a freelance writer for *The Union Democrat* newspaper for 13 years.

Thomas Lambert was born and raised in the Midwest, USA. A former U.S. Marine and Desert Storm veteran, he studied Creative Writing & Philosophy at The Universities of Kansas and East Anglia. His poetry has been featured in *Pearl, Di-Verse-City, Bluing The Blade, Beyond Words, The American Dissident* and other publications. Lambert lives in Dripping Springs, Texas with two daughters.

Mark Madigan is the author of a chapbook, *Thump and other poems*, published in 2019 by Finishing Line Press. An Army brat, he served for 29 years as DoD civilian providing intelligence and policy support for military operations before becoming a full-time writer.

LeeAnn Meadows, born and raised in Humboldt County, California, now calls New Mexico home. She lives on the outskirts of Las Cruces with her artist/husband Glenn Schwaiger and two dogs in an old adobe motel surrounded by pecan trees. Her poetry draws inspiration from nature, the human body, and relationships and has appeared in *Sin Fronteras, Adobe Walls, Malpais Review, Spiral Orb* and *CaliFragile.*

Louise Moises, an accomplished award-winning, published poet, frequently writes about family. Her father, brother and two husbands were all proud Navy men, veterans of three different wars. She particularly remembers the stories told by her father of when he was stationed at Moffet Field where the "Lighter Than Air" blimp fleets were docked and assigned to guard the coast. This poem is a memory shared by her parents.

Tom Nawrocki served in the U.S. Marine Corps from 1966-1970. He currently is an Associate Professor at Columbia College Chicago and an advisor for the Student Veterans Association.

Jonathan Pessant is a Maine poet and Army veteran. He holds a Master of Fine Arts from the Stonecoast MFA Program. His poems appear in *Collateral Journal*, *Pedestal Magazine*, Slipstream Press, and others.

Nicole Powers is the first-born daughter of an Army master sergeant and a M.A.S.H. unit surgical nurse during the occupation of postwar Japan serving on base for Korean war casualties. The children of military personnel all over the world are called "Brats."

Brendan Praniewicz earned his MFA in creative writing from San Diego State in 2007 and has subsequently taught creative writing at San Diego colleges. He has had poetry published in *From Whispers to Roars*, *Tiny Seed Journal*, *That Literary Review*, and *The Dallas Review*. In addition, he received second place in a first-chapters competition in the *Seven Hills Review* Chapter Competition in 2019. He won first place in The Rilla Askew Short Fiction Contest in 2020.

W.J. Price served in the U.S. Army for thirteen years, deploying five times in multiple campaigns before being medically retired as a Staff Sergeant. In addition to his service, W.J. Price has been a census-taker, a writing tutor, and is currently an adjunct professor of English at the Wentworth Institute of Technology.

H.V. Rhodes served in the U.S. Navy and is a life-long student of the American adventure. After a 24-year federal career as an engineer at Cape Canaveral, he retired and now serves as a librarian at the local Veterans' Memorial

Center. His Civil War Naval adventure novel *August 1864*, a SilverStowe Book, was published by and is available at The Write Thought Inc (www.TheWriteThought.com) as well as Amazon.com.

Jennifer McKeen Rodrigues (she/her) currently lives on the sacred Powhatan land of Fairfax, VA. She is trained as a certified yoga therapist & trauma informed yoga teacher, is a military spouse, & mom. She has been published in *The Muleskinner Journal, tiny frights, Military Experience & the Arts,* and *Susurrus Journal* among many other lovely journals as either poet or photographer.

Ricardo Ruiz is a multi-dimensional writer of poetry and prose. The son of potato factory workers, Ricardo hails from Othello, Washington. His work draws from his experience as a first-generation Mexican-American, and from his military service. Ricardo holds a Bachelor of Arts in Creative Writing from the University of Washington. His collection *We Had Our Resons* (Pulley Press) won the 2023 Washington State Book Award in poetry. While in the military, Ricardo earned the rank of Staff Sergeant while serving on four deployments, two to Afghanistan as an Infantryman 11B.

Patty Somlo's books, *Hairway to Heaven Stories* (Cherry Castle Publishing), *The First to Disappear* (Spuyten Duyvil) and *Even When Trapped Behind Clouds: A Memoir of Quiet Grace* (WiDo Publishing), have been Finalists in the International Book, Best Book, National Indie Excellence, American Fiction and Reader Views Literary Awards.

Last year, **Ben Stewart** sat down to write stories. Despite bothersome logistics regarding meals and subsequent bathroom breaks, he hasn't stopped since. His story "Young and Able-bodied" recently took second place in The President's Award for Prose at *Voices Magazine*. Before that, his story "To Good Health" was selected as a winner of a 250 word story contest in *Beyond Words Literary Magazine.*

Bruce Sydow is a combat-decorated Marine who served as a 19-year-old helicopter door gunner in Vietnam. He was President of the Faculty and Professor of the Year at Tacoma Community College and is the recipient of the Gold Star Award from the Sexual Assault Prevention Center of Pierce County. Bruce lives in a seaside village in Washington State on traditional land of the Coast Salish People with his wife, the artist and professor Danella Sydow.

Danella Sydow holds a Master of Fine Arts in Studio Art from Washington State University. She has served on the faculties of Olympic College, Pierce College, and Washington State University. The subject of the portrait painting "A 19-Year-Old Marine in Vietnam," is her husband, Bruce Sydow, a combat-decorated Marine and retired professor of social psychology. They will celebrate their 25th anniversary in 2025 and are planning a trip around the world to commemorate the milestone.

Kait Walser lives in New York City, where she hosted poetry readings and workshops. She won the 2013 Etruscan Press Prize at Wilkes University (MFA, 2014). You can find her poetry online and anthologized in *Everyday Escape Poems* (2015) and *In Absentia: Reflections on the Pandemic* (2020). A former U.S. Naval Sea Cadet, Kait is the proud daughter of a Navy veteran who worked for the VA and served as both a Corpsman and a Seabee.

While serving his 22-year military career, and again while earning his MFA from the University of Tampa, **Ben White** thought he was a poet. But—silly Ben—he is not a poet at all. Ben is a witness. What he writes is testimony. He is the author of *Conley Bottom: A Poemoir*, *The Recon Trilogy +1*, and *Always Ready: Poems from a Life in the U.S. Coast Guard.*

Irene T. Winslow is a retired kindergarten teacher and Women's Army Corps veteran. She enjoys walking in the park, reading mysteries and thrillers, and learning new crafts. She lives in Cleveland, Ohio.

Editor **Mary Senter** is a multi-disciplinary artist who creates in a cabin in the woods on the shores of the Salish Sea. She served in the U.S. Air Force as a structural specialist (CE) and as a DoD civilian carpenter/metalworker where she was honored as a Trailblazer by the New Mexico Commission on the Status of Women. She worked blue-collar jobs for years before returning to school, eventually earning certificates in literary fiction writing and an M.A. in strategic communication. She served as the graphic designer for *Crab Creek Review* and works in communications and design for local government. Her stories, essays, photos, and illustrations can be found in *North American Review, Gulf Stream, Drunk Monkeys, Ponder Review, Cleaver,* and elsewhere. She is the founder of Milltown Press. This is her first anthology.